AF480866

Unification of the Subluminal and the Superluminal in Cosmos Theory

Superluminal Parts of Leptons and Quarks
Faster Than Light Quantum Theory
Bradyon-Tachyon Based ElectroWeak and Strong Interaction Theories
Possibility of Superluminal Starship Motion

Stephen Blaha Ph. D.
Blaha Research

Pingree-Hill Publishing
MMXXIV

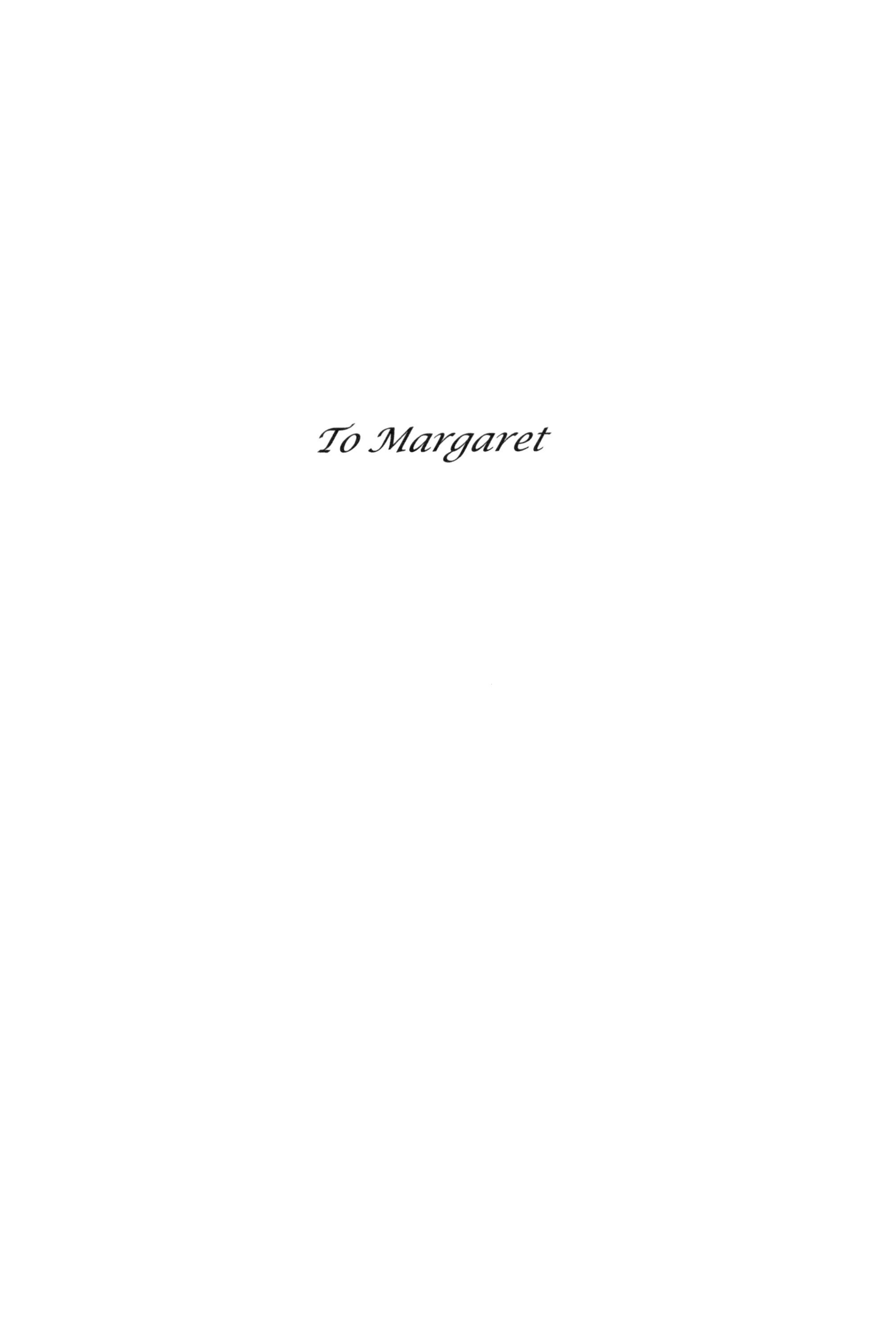

To Margaret

Some Other Books by Stephen Blaha

SuperCivilizations: Civilizations as Superorganisms (McMann-Fisher Publishing, Auburn, NH, 2010)

All the Universe! Faster Than Light Tachyon Quark Starships & Particle Accelerators with the LHC as a Prototype Starship Drive Scientific Edition (Pingree-Hill Publishing, Auburn, NH, 2011).

Unification of God Theory and Unified SuperStandard Model THIRD EDITION (Pingree Hill Publishing, Auburn, NH, 2018).

The Exact QED Calculation of the Fine Structure Constant Implies ALL 4D Universes have the Same Physics/Life Prospects (Pingree Hill Publishing, Auburn, NH, 2019).

Passing Through Nature to Eternity ProtoCosmos, HyperCosmos, Unified SuperStandard Theory (Pingree Hill Publishing, Auburn, NH, 2022).

HyperCosmos Fractionation and Fundamental Reference Frame Based Unification: Particle Inner Space Basis of Parton and Dual Resonance Models (Pingree Hill Publishing, Auburn, NH, 2022).

God and and Cosmos Theory (Pingree Hill Publishing, Auburn, NH, 2023).

Newton's Apple is Now The Fermion (Pingree Hill Publishing, Auburn, NH, 2023).

Cosmos Theory: The Sub-Particle Gambol Model (Pingree Hill Publishing, Auburn, NH, 2023).

Cosmos-Universe-Particle-Gambol Theory (Pingree Hill Publishing, Auburn, NH, 2024).

Fractal Cosmos Curve: Tensor-based Cosmos Theory (Pingree Hill Publishing, Auburn, NH, 2024).

The Eternal Form of Cosmos Theory Third Edition (Pingree Hill Publishing, Auburn, NH, 2024).

Geometric Cosmos Geometric Universe (Pingree Hill Publishing, Auburn, NH, 2024).

Particles and Universes of Cosmos Theory (Pingree Hill Publishing, Auburn, NH, 2024).

Available on Amazon.com, bn.com Amazon.co.uk and other international web sites as well as at better bookstores.

CONTENTS

FIGURES and TABLES

Introduction

The interest of this author in superluminal (faster than light) motion has been evident for many years. This book develops the dynamics of superluminal motion at the quantum level and then proceeds to generalize ElectroWeak Theory and Strong Interaction Theory to incorporate faster than light motion within them using new Duplex and Quadplex forms of quantum field theory developed by the author.

Duplex and Quadplex theories combine sublight and superluminal motion within individual quantum fields. These theories avoid the pitfalls of earlier efforts to understand faster than light motion. Particles are stable against drops into the vacuum by having a "normal" (sublight) part that prevents collapse. They have a tachyonic (faster than light) part for internal faster than light motion. There are no anomalies. One may "see" earlier times but one can't influence them. Time travel is not supported.

The possibility that the internal tachyon part may transform a particle to superluminal through interactions is explored.

Sublight (bradyon) and tachyon dynamics are examined in great detail. Then Duplex and Quadplex quantum field theories are developed to create new ElectroWeak and Strong Interaction Theories. These theories unite easily with our four dimension space-time using a quadplex formalism with four real-valued 4-dimension coordinate systems. One coordinate system for space-time; one for ElectroWeak interactions; and two for SU(4) Strong Interactions (that form a complex 4-dimension fundamental SU(4) representation.)

The origin of these new unified theories of ElectroWeak and Strong Interactions is in Cosmos Theory where their representation dimensions are adjacent in our universe's array of dimensions. Their form suggests the unification of these interactions.

These theories provide a new view of the internals of fundamental fermion particles and fields. They are naturally compatible within Cosmos Theory. They provide a new quantum view of the internal characteristics of fermions in ElectroWeak theory.

The book concludes with the development of a theory of total *Conserved* complex momentum in which the total real and total imaginary momentum parts are not separately conserved. Due this development, one can envision starships that would be capable of faster than light speeds using engines that generate complex momentum thrust. Such an engine remains for the future to develop.

1. Dimension and Quantum Fields

1.1 The Nature of Dimension

In developing the Cosmos Theory we considered space-time dimensions and internal symmetry dimensions. The fractal view of Cosmos Theory led us to consider growing dimensions from a zero dimension point to a two dimension grid.

The concept of dimension raises a number of points:

1. Dimensions do not have numerical values although we may assign labels to them in computations.
2. Dimensions cannot be numerically combined.
3. Things can combine dimensions. Cosmos spaces combine dimensions to specify space-times and internal symmetry groups. Concrete objects may have dimensions. For example, a rock.
4. Sets of dimensions may be specified by enumerating objects. For example, in Cosmos Theory we use the total number of components of the dimension array to specify the total number of dimensions of a space. Also, we specify the number of dimensions of the space-time of a space.

These facts are significant when one tries to understand the implications of the Unified SuperStandard Theory (UST) implied by Cosmos Theory. We find there is a grouping of dimensions due to nesting in the UST's dimension array. In particular, the Lorentz group $SO^+(1, 3)$ is combined with the ElectroWeak $SU(2) \otimes U(1)$ group in eight real dimensions with four dimensions allocated to each. This association of groups raises interesting questions:

1. Can one view the space of the $SU(2) \otimes U(1)$ group as a coordinate space similar to our space-time coordinate space?
2. What sort of space is it?
3. How do ElectroWeak particles arise within this context?
4. What sort of Quantum Field Theory would embody both coordinate systems?
5. Would the Quantum Field Theory support a bradyon or a tachyon formulation?.

This book proposes to define quantum field coordinate system wave functions where one coordinate system (*external* to particles) has a bradyon structure and deals with conventional space-time. The other coordinate system will be *internal* to a particle—a tachyon implementation in imaginary coordinates. We will then associate the internal wave functions with ElectroWeak isodoublets.

The proposed association of the Lorentz group $SO^+(1, 3)$ with the ElectroWeak $SU(2) \otimes U(1)$ group in eight real dimensions raises the question of a further association of these groups with the eight real dimension broken SU(4) group. Is there a further association in a yet bigger combined space? We answer those questions in the following chapters.

The origin of these new unified theories of ElectroWeak and Strong Interactions is in Cosmos Theory where their representation dimensions are adjacent in our universe's Cosmos dimension array. See Fig. 1.1. Their form suggests the unification of the interactions.

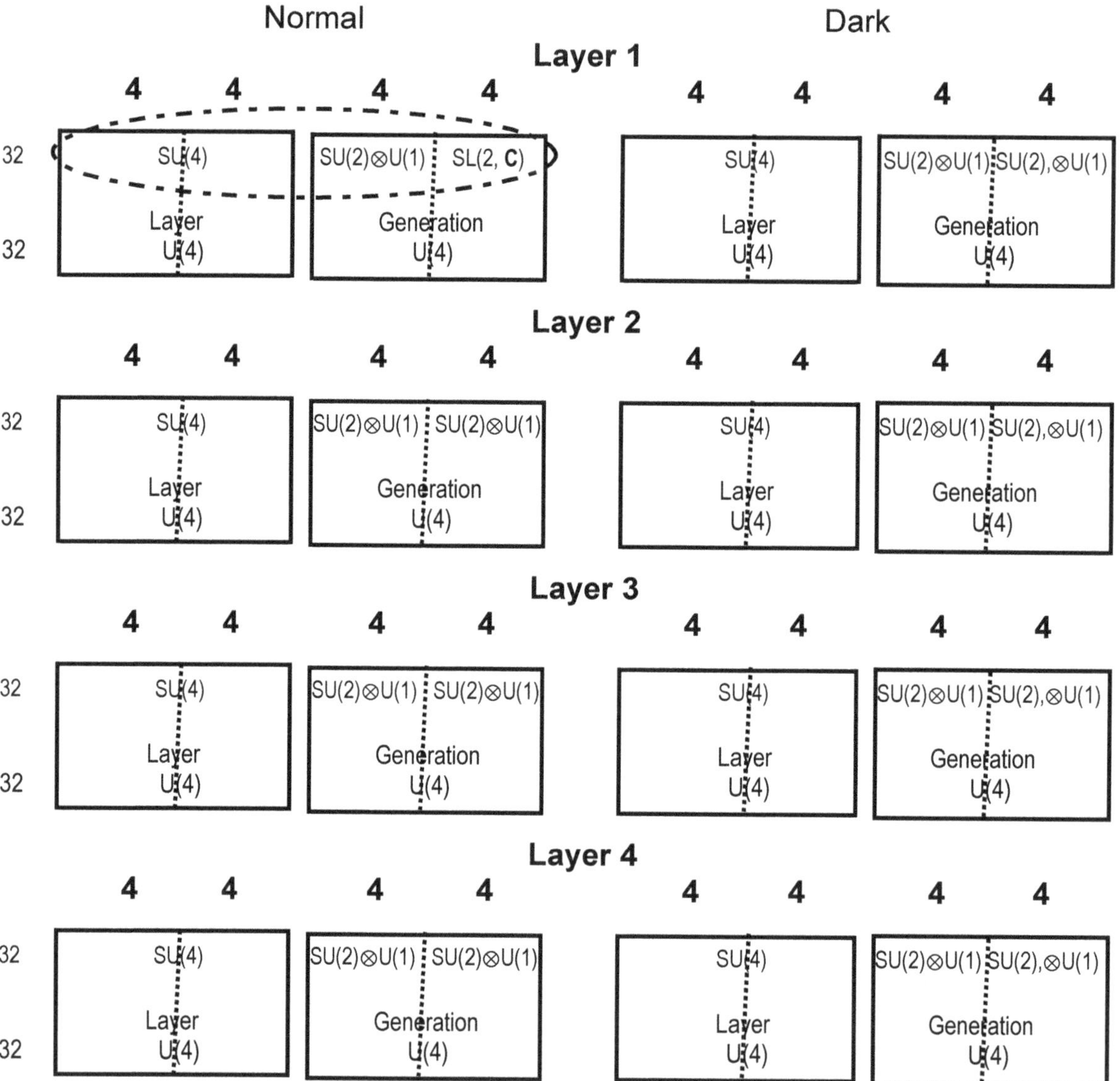

Figure 1.1. from Blaha (2020d). Normal and Dark symmetry groups of UST. SL(2, C) represents the Lorentz group SO+(1,3). Note the 16 dimensions of SU(4), SU(2)⊗U(1), and SL(2, C) are adjacent in Layer 1 suggesting a unification possibility.

2. Overview of New QFT Formalism for ElectroWeak and Strong Interactions

2.1 A More Detailed View

This book provides a new formalism for The Unified SuperStandard Theory's (UST) Lorentz group $SO^+(1, 3)$ and its ElectroWeak $SU(2) \otimes U(1)$ and Strong Interaction $SU(4)$ symmetries. *The starting point is the grouping of $SU(2) \otimes U(1)$ and $SO^+(1, 3)$ in the Cayley-Dickson $n = 3$ dimension array* for our universe where $SO^+(1,3)$ has the four dimension coordinate system denoted y and $SU(2) \otimes U(1)$ has a real four dimension coordinate system denoted z. Then $SU(4)$ is introduced to create a unified set of interactions. See Fig. 1.1. $SU(4)$ has two real four dimension coordinate systems denoted u and v supporting an $SU(4)$ complex fundamental representation. The book establishes unified representations of these symmetries by considering the fundamental representations of these groups.

We will show that the $SU(2) \otimes U(1)$ four dimension fundamental representation is a representation that may be combined with the four dimension space-time of $SO^+(1, 3)$ to form a complex space-time. This combined space-time supports quantum fields that we call Duplex Fermion Fields. One field, the external field, is a bradyon field. The other field is internal and is either a bradyon field or tachyon field.

We proceed through the following steps:

1. Show the PseudoFermion fields form support $SO^+(1, 3)$ and $SU(2) \otimes U(1)$.

2. Show $SU(4)$ may be similarly treated using a four complex dimension fundamental representation consisting of two real four dimension coordinate systems. See Fig. 1.1.

3. Show that a combined theory results consisting of four separate real-valued sets of coordinates: one four real coordinates set for $SO^+(1,3)$, one four real coordinates set for $SU(2) \otimes U(1)$, one four real coordinates set for the real part of the $SU(4)$ representation, and one four real coordinates set for the imaginary part of the $SU(4)$ representation.

In chapter 6 we implement a complex space-time for a real four dimension representation of $SO^+(1, 3)$, *which is implicit in the y coordinates*, and a real four dimension representation of $SU(2) \otimes U(1)$ in the z coordinates. PseudoFermion Quantum fields in this combined representation may be defined. These fields are two coordinate system fields with y and z coordinates. A free fields Lagrangian for this case has the quantum field form:

$$\mathscr{L} = \overline{\psi}^{\alpha\beta}[iM^{-1}\gamma_{y\alpha\kappa}{}^{\mu}\partial/\partial y^{\mu}\ \gamma_{z\beta\lambda}{}^{\nu}\partial/\partial z^{\nu} - M]\psi^{\kappa\lambda} \tag{2.1}$$

where $\psi^{\kappa\lambda}(y, z)$ is a normal Quantum field, $\gamma_y{}^{\mu}$ and $\gamma_z{}^{\mu}$ are Dirac matrices for the y and z coordinates respectively, and y and z are coordinates in r dimension space-times, and M is the mass. Chapter 5 details the resulting field theory.

In chapter 6 we implement a space-time for a four real dimension representation of $SO^+(1, 3)$ denoted y, and a four real dimension representation coordinate system for $SU(2)\otimes U(1)$ denoted z. In chapter 9 we then define a four real dimension representation for the real part of the complex SU(4) representation denoted u, and a four real dimension representation for the imaginary part of the complex SU(4) representation denoted v.

2.2 $SO^+(1, 3)\otimes SU(2)\otimes U(1)$ Free Field Case

PseudoFermion Quantum fields may be defined for two coordinate systems using y and z coordinates. A free fields Lagrangian for this case has the form:

$$\begin{aligned}
\mathscr{L} = \overline{\psi}_2{}^{\alpha\beta}[iM^{-1}\gamma_{y\alpha\kappa}{}^{\mu}\partial/\partial y^{\mu}\gamma_{z\beta\lambda}{}^{\nu}\partial/\partial z^{\nu} - M]\psi_1{}^{\kappa\lambda} + \\
+ \overline{\psi}_1{}^{\alpha\beta}[iM^{-1}\gamma_{y\alpha\kappa}{}^{\mu}\partial/\partial y^{\mu}\ \gamma_{z\beta\lambda}{}^{\nu}\partial/\partial z^{\nu} - M]\psi_2{}^{\kappa\lambda}
\end{aligned} \tag{2.2}$$

where $\psi_1{}^{\kappa\lambda}(y, z)$ and $\psi_1{}^{\kappa\lambda}(y, z)$ are PseudoQuantum fields, $\gamma_y{}^{\mu}$ and $\gamma_z{}^{\mu}$ are Dirac matrices for the y and z coordinates respectively, and y and z are coordinates in r dimension space-times, and M is the mass. Chapter 6 details the resulting field theory.

2.3 SU(4) Free Field Case

The complex SU(4) fundamental representation has a part for the real 4 dimension part and a part for the imaginary 4 dimension part. Both parts support bradyon and tachyon quantum fields. A free PseudoQuantum Lagrangian for the combined symmetries is

$$\begin{aligned}
\mathscr{L} = \overline{\psi}_2{}^{\alpha\beta\rho\sigma}[-M^{-3}\gamma_{y\alpha\kappa}{}^{\mu}\partial/\partial y^{\mu}\ \gamma_{z\beta\lambda}{}^{\nu}\partial/\partial z^{\nu}\ \gamma_{u\rho\tau}{}^{\mu}\partial/\partial u^{\mu}\ \gamma_{v\sigma\chi}{}^{\nu}\partial/\partial v^{\nu} - M]\psi_1{}^{\kappa\lambda\tau\chi} + \\
+ \overline{\psi}_1{}^{\alpha\beta\rho\sigma}[-M^{-3}\gamma_{y\alpha\kappa}{}^{\mu}\partial/\partial y^{\mu}\ \gamma_{z\beta\lambda}{}^{\nu}\partial/\partial z^{\nu}\ \gamma_{u\rho\tau}{}^{\mu}\partial/\partial u^{\mu}\ \gamma_{v\sigma\chi}{}^{\nu}\partial/\partial v^{\nu} - M]\psi_2{}^{\kappa\lambda\tau\chi}
\end{aligned} \tag{2.3}$$

where $\gamma_y{}^{\mu}$, $\gamma_z{}^{\mu}$, $\gamma_u{}^{\mu}$ and $\gamma_v{}^{\mu}$ are Dirac matrices for the y, z, u and v coordinates respectively, and M is the common mass.

The equations of motion are

$$[iM^{-3}\gamma_y{}^{\mu}\partial/\partial y^{\mu}\ \gamma_z{}^{\nu}\partial/\partial z^{\nu}\ \gamma_u{}^{\mu}\partial/\partial u^{\mu}\ \gamma_v{}^{\nu}\partial/\partial v^{\nu} - M]\psi_1 = 0 \tag{2.4}$$

$$[iM^{-3}\gamma_y{}^{\mu}\cdot\partial/\partial y^{\mu}\ \gamma_z{}^{\nu}\partial/\partial z^{\nu}\ \gamma_u{}^{\mu}\partial/\partial u^{\mu}\ \gamma_v{}^{\nu}\partial/\partial v^{\nu} - M]\psi_2 = 0 \tag{2.5}$$

We may define subsidiary equations of motion

$SO^+(1,3)$	Bradyon	$[i\gamma_y{}^\mu \cdot \partial/\partial y^\mu - M]\psi_j = 0$	(2.6)
$SU(2)\otimes U(1)$	Bradyon	$[i\gamma_z{}^\mu \cdot \partial/\partial z^\mu - M]\psi_j = 0$	
	or Tachyon	$[\gamma_z{}^\nu \partial/\partial z^\nu - M]\psi_j = 0$	
$SU(4)$ representation real part	Bradyon	$[i\gamma_u{}^\nu \partial/\partial u^\nu - M]\psi_j = 0$	
	or Tachyon	$[\gamma_u{}^\nu \partial/\partial u^\nu - M]\psi_j = 0$	
$SU(4)$ representation imaginary part	Bradyon	$[i\gamma_v{}^\nu \partial/\partial v^\nu - M]\psi_j = 0$	
	or Tachyon	$[\gamma_v{}^\nu \partial/\partial v^\nu - M]\psi_j = 0$	

The result of these efforts is a formulation of some UST symmetries that explains the somewhat unusual features of ElectroWeak theory and SU(4). See chapter 9 for details.

The formulation avoids supposed problems of faster than light phenomena. It opens the door to the consideration of faster than light travel.

3. Bradyon and Tachyon Quantum Field Theories

The development of Quantum Field Theories, needed to address the issues raised in chapters 1 and 2, requires formalisms for bradyon (below the speed of light) quantum fields and tachyon (above the speed of light) quantum fields. These types of quantum fields are discussed in this chapter and its appendix. They are abstracted, with some changes, from Blaha (2018e).

The following chapters consider a new formalism based on combining PseudoFermion wave functions with differing external and internal parts coordinate systems. The set of wave functions contain both bradyon and tachyon parts in general.

Tachyon fields are initially restricted to particle interiors eliminating one supposed issue with tachyons.[1] This chapter eliminates another issue: the momentum limitation of tachyons that conflicts with canonical anti-commutation relations. This author developed a *canonical light-front tachyon formulation* that resolves this issue. (See section 3.4.4.2 below and Blaha (2018e), and earlier books.) Thus shortcomings found in previous efforts using tachyons are resolved.

3.1 Matrix Representation of Complex Lorentz Group L_C Boosts

The remainder of this chapter is based on the Complex Lorentz group. We begin with Complex Lorentz Group (L_C) boosts because they will be crucial in the determination of the equations of motion of various types of spin ½ particles. An L_C boost can be expressed in the form

$$\Lambda_C(\mathbf{v_c}) \equiv \Lambda_C(\omega, \hat{\mathbf{w}}) \qquad (3.1)$$

$$\Lambda_C(\mathbf{v_c}) = \exp[i\omega\hat{\mathbf{w}}\cdot\mathbf{K}] \qquad (3.2)$$

where

$$\omega = (\omega_r^2 - \omega_i^2 + 2i\omega_r\omega_i\,\hat{\mathbf{u}}_r\cdot\hat{\mathbf{u}}_i)^{\frac{1}{2}} \qquad (3.3)$$

and

$$\hat{\mathbf{w}} = (\omega_r\hat{\mathbf{u}}_r + i\omega_i\hat{\mathbf{u}}_i)/\omega \qquad (3.4)$$

Since $\hat{\mathbf{u}}_r\cdot\hat{\mathbf{u}}_r = 1 = \hat{\mathbf{u}}_i\cdot\hat{\mathbf{u}}_i$

$$\hat{\mathbf{w}}\cdot\hat{\mathbf{w}} = 1 \qquad (3.5)$$

and the complex relative velocity is

$$\mathbf{v_c} = \hat{\mathbf{w}}\,\tanh(\omega) \qquad (3.6)$$

We now analytically continue to complex ω and complex unit vectors $\hat{\mathbf{w}}$. The resulting complex generalization will be the matrix form of proper L_C boosts:

---

[1] No causality violation at large distances outside of particles and interaction regions.

$$\Lambda_C(\mathbf{v_c}) = \exp[i\omega\hat{\mathbf{w}}\cdot\mathbf{K}] \equiv \Lambda_C(\omega, \hat{\mathbf{w}})$$

$$= \begin{bmatrix} \cosh(\omega) & -\sinh(\omega)\hat{w}_x & -\sinh(\omega)\hat{w}_y & -\sinh(\omega)\hat{w}_z \\ -\sinh(\omega)\hat{w}_x & 1 + (\cosh(\omega) - 1)\hat{w}_x^2 & (\cosh(\omega) - 1)\hat{w}_x\hat{w}_y & (\cosh(\omega) - 1)\hat{w}_x\hat{w}_z \\ -\sinh(\omega)\hat{w}_y & (\cosh(\omega) - 1)\hat{w}_x\hat{w}_y & 1 + (\cosh(\omega) - 1)\hat{w}_y^2 & (\cosh(\omega) - 1)\hat{w}_y\hat{w}_z \\ -\sinh(\omega)\hat{w}_z & (\cosh(\omega) - 1)\hat{w}_x\hat{w}_z & (\cosh(\omega) - 1)\hat{w}_y\hat{w}_z & 1 + (\cosh(\omega) - 1)\hat{w}_z^2 \end{bmatrix} \quad (3.7)$$

Since analytic continuations are unique, the above form for $\Lambda_C(\mathbf{v_c})$ is well-defined and unique. It spans the complete set of proper L_C boosts.

3.2 Left-handed and Right-handed Parts of L_C

We now describe the Left-handed and Right-handed parts[2] of L_C boosts.

3.2.1 Left-handed Part of L_C

If we let

$$\hat{\mathbf{u}}_i = \hat{\mathbf{u}}_r \equiv \hat{\mathbf{u}} \quad (3.8)$$

so that the vector $\hat{\mathbf{u}}_i$ is parallel to $\hat{\mathbf{u}}_r$, and

$$\omega_i = \pi/2 \quad (3.9)$$

then $\Lambda_C(\mathbf{v_c})$ becomes a Left-handed L_C boost:

$$\Lambda_C(\mathbf{v_c}) = \Lambda_L(\omega_r, \mathbf{u}) \quad (3.10)$$

3.2.2 Right-handed part of L_C

If we let

$$\hat{\mathbf{u}}_i = -\hat{\mathbf{u}}_r \equiv -\hat{\mathbf{u}} \quad (3.11)$$

so that the vector $\hat{\mathbf{u}}_i$ is anti-parallel to $\hat{\mathbf{u}}_r$, and

$$\omega_i = -\pi/2 \quad (3.12)$$

then $\Lambda_C(\mathbf{v_c})$ becomes a Right-handed L_C boost:

$$\Lambda_C(\mathbf{v_c}) = \Lambda_R(\omega_r, \mathbf{u}) \quad (3.13)$$

as described in Blaha (2007b).

3.3 Difference between the Parts of L_C Reduced to Parallelism of $\hat{\mathbf{u}}_r$ and $\hat{\mathbf{u}}_i$

Since the Left-handed L_C part leads to the Standard Model's left-handed features, it seems that the parallel case $\hat{\mathbf{u}}_i = \hat{\mathbf{u}}_r \equiv \hat{\mathbf{u}}$ is more favored by Nature.[3] To some

---

[2] The designations Left-handed and Right-handed are chosen to reflect the Left-handed and Right-handed fermion fields that will be used to construct The Standard Model later. See Blaha (2007b) for more detail.

extent this concept of parallel vectors $\hat{u}_i$ and $\hat{u}_r$, which leads to the Left-handed L_C, is more intuitively satisfying than the anti-parallel case that leads to the Right-handed L_C part. However, a deeper reason for Nature's choice remains to be found.

3.4 Free Spin ½ Particles – Leptons & Quarks

In this section we begin by developing dynamical equations for spin ½ particles based on the L_C parts. These spin ½ particles are conventional Dirac particles (Majorana particles are also allowed but not discussed), spin ½ tachyons, and "color" versions of both types. We will identify leptons and quarks with these fields.

3.4.1 Introduction

Tachyons are particles that move faster than the speed of light. As we saw in earlier books tachyons exist inside Black Holes, and within current theories – particularly SuperString theories. There are also experimental indications that neutrinos are tachyons.

Attempts to create canonical tachyon quantum field theories began in the 1960's. These attempts were made within the framework of the Lorentz group and, consequently, were limited to spin 0 theories since there are no finite dimensional representations of the Lorentz group for negative m^2 except for the one-dimensional representation. None of these attempts, or attempts since then, succeeded in creating an acceptable canonically quantized spin 0 tachyon quantum field theory.[4]

In this section we will formulate a free spin ½ tachyon[5] Quantum Field Theory. We choose to develop a normal spin ½ theory first. Then we develop a free spin ½ tachyon theory because, as we will see, spin ½ tachyon particles (quarks and leptons) play an extraordinary role in our new ElectroWeak Model.

We will develop our spin ½ tachyon theory from the "ground up" by applying a Left-Handed L_C boost to the Dirac equation, and its Dirac spinor wave function, for a particle at rest. This procedure will give a tachyon spinor wave function, and the momentum space tachyon equation equivalent of the Dirac equation. Then we will obtain the coordinate space tachyon Dirac equation, define a Lagrangian, and proceed to create a canonical quantum field theory for spin ½ tachyons.

3.4.2 First Step - Deriving the Conventional Dirac Equation

In this section we will review a method of obtaining the equation of motion of a particle using a free Dirac equation that is obtained by a Lorentz boost of a spinor wave function[6] of a particle at rest.

In the case of a Lorentz transformation the 4×4 matrix form of a Lorentz transformation of Dirac matrices is

---

[3] It is possible that parity violation might disappear at ultra-high energies. Then we would view the parity symmetric theory as broken to the left-handed Standard Model currently established by experiment with right-handed parts at higher energy.

[4] Except Blaha (2006).

[5] It differs significantly from tachyon theories such as those of G. Feinberg and E. C. Sudarshan.

[6] The spinor wave function of a particle at rest is a 4-vector of the 4×4 matrix representation of 4-valued Asynchronous Logic.

$$S^{-1}(\Lambda(v))\gamma^{\nu}S(\Lambda(v)) = \Lambda^{\nu}{}_{\mu}(v)\gamma^{\mu} \tag{3.14}$$

where $S(\Lambda(v))$ is

$$S(\Lambda(v)) = \exp(-i\omega\sigma_{0i}v_i/(2|\mathbf{v}|)) = \exp(-\omega\gamma^0\gamma\cdot\mathbf{v}/(2|\mathbf{v}|))$$
$$= \cosh(\omega/2)I + \sinh(\omega/2)\gamma^0\gamma\cdot\mathbf{p}/|\mathbf{p}| \tag{3.15}$$

with $\omega = \text{arctanh}(|\mathbf{v}|)$, $\cosh(\omega/2) = [(E+m)/(2m)]^{\frac{1}{2}}$ and $\sinh(\omega/2) = |\mathbf{p}|[2m(E+m)]^{-\frac{1}{2}}$. Also

$$S^{-1}(\Lambda(v)) = \gamma^0 S^{\dagger}(\Lambda(v))\gamma^0 = \exp(\omega\gamma^0\gamma\cdot\mathbf{v}/(2|\mathbf{v}|))$$
$$= \cosh(\omega/2)I - \sinh(\omega/2)\gamma^0\gamma\cdot\mathbf{p}/|\mathbf{p}| \tag{3.16}$$

In constructing fermion dynamical equations *we shall assume that they are linear in derivatives* (although a quadratic form is possible.) We will use the sixteen 4×4 matrices that span the set of transformations of the four values of Asynchronous Logic. Since, by theorem,[7] all 4×4 γ matrices are equivalent up to a unitary transformation we can rotate any constant matrix into a multiple of γ^0 without loss of generality.

We begin by defining a generic positive energy plane wave solution of the Dirac equation for a normal fermion particle at rest with rest mass m, *which we take to be the bare mass in the absence of interactions*,[8] as

$$\psi(x) = e^{-imt}w(0) \tag{3.17}$$

with $w(0)$ a four component logic spinor column vector. *For a free particle at rest, the rest energy $m = m_0$, the mass.* The wave function satisfies the momentum space Dirac equation for a fermion at rest:

$$(m\gamma^0 - m)e^{-imt}w(0) = 0 \tag{3.18}$$

Subsequently we will use a similar procedure to construct the free tachyonic Dirac equation.

If we now apply $S(\Lambda(v))$ we find

$$0 = S(\Lambda(v))(m\gamma^0 - m)e^{-imt}w(0) = [mS(\Lambda(v))\gamma^0 S^{-1}(\Lambda(v)) - m]S(\Lambda(v))w(0)$$

A straightforward evaluation shows

$$mS(\Lambda(v))\gamma^0 S^{-1}(\Lambda(v)) = g_{\mu\nu}p^{\mu}\gamma^{\nu} = \not{p} \tag{3.19}$$

where $p^0 = (p^2 + m^2)^{\frac{1}{2}}$, $\mathbf{p} = \gamma m\mathbf{v}$, and $p = |\mathbf{p}|$. In addition

---

[7] R. H. Good, Rev. Mod. Phys., **27**, 187 (1955).

[8] As state earlier the derivation proceeds in steps from the Complex Lorentz group to free fermions and thence to interacting fermions. **We use the bare mass throughout the free fermion discussions in this chapter. $m = m_0$.**

$$S(\Lambda(v))w(0) = w(p) \tag{3.20}$$

is a positive energy Dirac spinor. Therefore the Dirac equation for a fermion in motion in momentum space has the form:

$$(\not{p} - m)e^{-ip\cdot x}w(p) = 0 \tag{3.21}$$

where the exponential factor, mt, is also boosted to p·x. Eq. 3.21 implies the well-known free, coordinate space Dirac equation:

$$(i\gamma^\mu \partial/\partial x^\mu - m)\psi(x) = 0 \tag{3.22}$$

3.4.3 Derivation of the Tachyon Dirac Equation

The Left-handed boost has the form:

$$\Lambda_L(\omega, \mathbf{u}) = \Lambda(\omega + i\pi/2, \mathbf{u}) = \exp[i\omega_L \hat{\mathbf{u}} \cdot \mathbf{K}] \tag{3.23}$$

where

$$\omega_L = \omega + i\pi/2 \tag{3.23a}$$

and

$$\cosh(\omega_L) = i\sinh(\omega) = -\gamma = i\gamma_s \tag{3.24}$$
$$\sinh(\omega_L) = i\cosh(\omega) = -\beta\gamma = i\beta\gamma_s$$

with, $\beta = v > 1$, $\gamma_s = (\beta^2 - 1)^{-\frac{1}{2}}$, and $\omega \geq 0$. The rapidity w is

$$w = \text{artanh}(\omega_L) = \beta \tag{3.24a}$$

Thus

$$\sinh(\omega) = \gamma_s \tag{3.25}$$
$$\cosh(\omega) = \beta\gamma_s$$

and

$$w^{-1} = \text{artanh}(\omega) = \beta^{-1} \tag{3.24a}$$

The corresponding spinor transformation is:

$$S_L(\Lambda_L(\omega, \mathbf{u})) = \exp(-i\omega_L\sigma_{0i}v_i/(2|\mathbf{v}|)) = \exp(-\omega_L\gamma^0\boldsymbol{\gamma}\cdot\mathbf{v}/(2|\mathbf{v}|))$$
$$= \cosh(\omega_L/2)I + \sinh(\omega_L/2)\gamma^0\boldsymbol{\gamma}\cdot\mathbf{p}/|\mathbf{p}| \tag{3.26}$$

The inverse transformation is

$$S_L^{-1}(\Lambda_L(\omega, \mathbf{u})) = \gamma^2\gamma^0 K^{-1}S_L^\dagger K\gamma^0\gamma^2 = \gamma^2\gamma^0 S_L^{\ T}\gamma^0\gamma^2 = \exp(\omega_L\gamma^0\boldsymbol{\gamma}\cdot\mathbf{v}/(2|\mathbf{v}|))$$
$$= \cosh(\omega_L/2)I - \sinh(\omega_L/2)\gamma^0\boldsymbol{\gamma}\cdot\mathbf{p}/|\mathbf{p}| \tag{3.27}$$

where the superscript T denotes the transpose and K is the complex conjugation operator (that also appears in the time-reversal operator). Note that S_L is not unitary just as the equivalent spinor Lorentz transformation $S(\Lambda(v))$ is not unitary.

We can now apply a left-handed superluminal transformation to the generic positive energy plane wave solution of the Dirac equation for a particle of mass m at rest. The result is

$$0 = S_L(\Lambda_L(\omega, \mathbf{u}))(m\gamma^0 - m)e^{-imt}w(0)$$
$$= [mS_L\gamma^0 S_L^{-1} - m]e^{-imt}S_Lw(0)$$

where $S_L = S_L(\Lambda_L(\omega, \mathbf{u}))$. After some algebra

$$mS_L\gamma^0 S_L^{-1} = m[\cosh(\omega_L)\gamma^0 - \sinh(\omega_L)\boldsymbol{\gamma}\cdot\mathbf{p}/|\mathbf{p}|]$$
$$= i\gamma^0 E - i\boldsymbol{\gamma}\cdot\mathbf{p} = i\not{p} \qquad (3.28)$$

using the tachyon energy and momentum expressions

$$\mathbf{p} = mv\gamma_s \qquad\qquad E = m\gamma_s \qquad (3.29)$$

Also

$$S_Lw(0) = w_T(p) \qquad (3.30)$$

is a tachyon spinor. See Appendix 3-A for a discussion of tachyon spinors.

The momentum space tachyonic Dirac equation is

$$(i\not{p} - m)e^{ip\cdot x}w_T(p) = 0 \qquad (3.31)$$

where $p\cdot x = Et - \mathbf{p}\cdot\mathbf{x}$ after performing a corresponding left-handed superluminal coordinate transformation in the exponential factor. Thus a positive energy wave is transformed into a negative energy wave by the superluminal transformation.

If we apply $i\not{p}$ to we find the tachyon mass condition is satisfied

$$-E^2 + \mathbf{p}^2 = m^2 \qquad (3.32)$$

Transforming back to coordinate space we obtain the *tachyon Dirac equation*:

$$(\gamma^\mu\partial/\partial x^\mu - m)\psi_T(x) = 0 \qquad (3.33)$$

The "missing" factor of i in the first term of eq. 3.33 requires the Lagrangian to be different from the conventional Dirac Lagrangian in order for the Lagrangian to be real. The simplest, physically acceptable, free spin ½ tachyon Lagrangian density is:

$$\mathcal{L}_T = \psi_T^S(\gamma^\mu\partial/\partial x^\mu - m)\psi_T(x) \qquad (3.34)$$

where

$$\psi_T{}^S = \psi_T{}^\dagger \, i\gamma^0\gamma^5 \tag{3.35}$$

The corresponding action is

$$I = \int d^4x \, \mathcal{L}_T \tag{3.36}$$

Appendix 3-B of Blaha (2007b) proves I is real. The Hamiltonian density is

$$\mathcal{H} = \pi_T \dot{\psi}_T - \mathcal{L} = i\psi_T{}^\dagger \gamma^5 (\boldsymbol{\alpha}\cdot\nabla + \beta m)\psi_T = -i\psi_T{}^\dagger \gamma^5 \dot{\psi}_T \tag{3.37}$$

using the tachyon Dirac equation to obtain the last equality. The reader will note that the tachyon Hamiltonian is Hermitean by explicit calculation up to an irrelevant total spatial divergence.

3.4.3.1 Probability Conservation Law

The tachyon Dirac equation implies a probability conservation law:

$$\partial\rho_5/\partial t = \nabla\cdot\mathbf{j}_5 \tag{3.38}$$

where

$$\rho_5 = \psi_T{}^\dagger \gamma^5 \psi_T \qquad\qquad \mathbf{j}_5 = \psi_T{}^\dagger \gamma^5 \boldsymbol{\alpha}\psi_T \tag{3.39}$$

We are thus led to define the conserved axial charge Q_5

$$Q_5 = \int d^3x \, \psi_T{}^\dagger \gamma^5 \psi_T \tag{3.40}$$

3.4.3.2 Energy-Momentum Tensor

The tachyon energy-momentum tensor is

$$\mathcal{T}_{T\mu\nu} = -\,g_{\mu\nu}\,\mathcal{L}_T + \partial\mathcal{L}_T/\partial(\partial\psi_T/\partial x_\mu)\,\partial\psi_T/\partial x^\nu \tag{3.41}$$
$$= i\psi_T{}^\dagger \gamma^0\gamma^5\gamma_\mu\partial\psi_T/\partial x^\nu \tag{3.42}$$

and thus the conserved energy and momentum are

$$P^0 = H = \int d^3x \, \mathcal{T}_T{}^{00} = i\int d^3x\psi_T{}^\dagger \gamma^5 (\boldsymbol{\alpha}\cdot\nabla + \beta m)\psi_T \tag{3.43}$$
$$P^i = \int d^3x \, \mathcal{T}_T{}^{0i} = -\,i\int d^3x \, \psi_T{}^\dagger \gamma^5 \partial\psi_T/\partial x_i \tag{3.44}$$

Both the energy and momentum differ significantly from the corresponding quantities for conventional Dirac fields.

3.4.4 Tachyon Canonical Quantization

Having defined a suitable tachyon Lagrangian we can now proceed to its canonical quantization. The conjugate momentum can be calculated from the above Lagrangian density:

$$\pi_{Ta} = \partial\mathcal{L}_T/\partial\dot{\psi}_{Ta} \equiv \partial\mathcal{L}_T/\partial(\partial\psi_{Ta}/\partial t) = -i(\psi_T{}^\dagger\gamma^5)_a \tag{3.45}$$

The resulting non-zero, canonical anti-commutation relations are

$$\{\pi_{Ta}(x), \psi_{Tb}(x')\} = i\,\delta_{ab}\,\delta^3(x - x')$$

$$\{\psi_T{}^\dagger{}_a(x), \psi_{Tb}(x')\} = -[\gamma^5]_{ab}\,\delta^3(x - x') \tag{3.46}$$

At this point we might attempt to complete the canonical quantization procedure in the conventional manner by Fourier expanding the quantum field and specifying anti-commutation relations for the Fourier component amplitudes. However the incompleteness of the set of plane waves, which are limited by the restriction $|p| \geq m$, causes the anti-commutator of the fields not to yield a $\delta^3(x - x')$. Thus the conventional approach fails to yield the required anti-commutation relations.[9]

Other approaches: 1) decompose the tachyon field into left-handed and right-handed parts and then second quantize each part; and 2) second quantize in light-front coordinates $(x^\pm = (x^0 \pm x^3)/\sqrt{2})$. These approaches also both fail.[10]

The only approach that does succeed[11] *is to decompose the tachyon field into left-handed and right-handed parts and then second quantize in light-front coordinates. We follow this procedure in the following subsections.*

3.4.4.1 Separation into Left-Handed and Right-Handed Fields

We will use a transformed set of Dirac matrices to develop our left-handed and right-handed tachyon formulations:

$$\gamma^0 = \begin{bmatrix} 0 & -I \\ -I & 0 \end{bmatrix} \qquad \gamma^i = \begin{bmatrix} 0 & \sigma_i \\ -\sigma_i & 0 \end{bmatrix} \qquad \gamma^5 = \begin{bmatrix} I & 0 \\ 0 & -I \end{bmatrix}$$

$$\tag{3.47}$$

which are obtained from the usual Dirac matrices by applying the unitary transformation $U = 2^{1/2}(I + \gamma^5\gamma^0)$. *I is the 4×4 identity matrix in eq. 3.47.* The γ^5 chirality operator's eigenvalues define handedness: +1 corresponds to right-handed; and −1 corresponds to left-handed:

$$\gamma^5\psi_L = -\psi_L \qquad\qquad \gamma^5\psi_R = \psi_R \tag{3.48}$$

---

[9] See G. Feinberg, Phys. Rev. **159**, 1089 (1967) for example.
[10] See the first edition Blaha (2006) where these possibilities were considered and found to fail.
[11] Blaha (2006) discusses this case in detail.

Consequently, we can define left-handed and right-handed tachyon fields with the projection operators:

$$C^{\pm} = \tfrac{1}{2}(I \pm \gamma^5)$$
$$C^{+} + C^{-} = I$$
$$C^{\pm\,2} = C^{\pm}$$
$$C^{+}C^{-} = 0$$

$$(3.49)$$

with the result

$$\psi_{TL} = C^{-}\psi_T$$
$$\psi_{TR} = C^{+}\psi_T$$

$$(3.50)$$

We can calculate the commutation relations of the left-handed and right-handed tachyon fields from eq. 3.46 by pre-multiplying and post-multiplying by $\tfrac{1}{2}(1 - \gamma^5)$ and $\tfrac{1}{2}(1 + \gamma^5)$. The results are:

$$\{\psi_{TLa}^{\dagger}(x),\ \psi_{TLb}(x')\} = \tfrac{1}{2}(1 - \gamma^5)_{ab}\,\delta^3(x - x')$$

$$(3.51)$$

$$\{\psi_{TRa}^{\dagger}(x),\ \psi_{TRb}(x')\} = -\tfrac{1}{2}(1 + \gamma^5)_{ab}\,\delta^3(x - x')$$

$$(3.52)$$

$$\{\psi_{TLa}^{\dagger}(x),\ \psi_{TRb}(x')\} = \{\psi_{TRa}^{\dagger}(x),\ \psi_{TLb}(x')\} = 0$$

$$(3.53)$$

The Lagrangian density above decomposes into left-handed and right-handed parts:

$$\mathcal{L}_T = \psi_{TL}^{\dagger}\gamma^0 i\gamma^{\mu}\partial_{\mu}\psi_{TL} - \psi_{TR}^{\dagger}\gamma^0 i\gamma^{\mu}\partial_{\mu}\psi_{TR} - im[\psi_{TR}^{\dagger}\gamma^0\psi_{TL} - \psi_{TL}^{\dagger}\gamma^0\psi_{TR}]$$

$$(3.54)$$

3.4.4.2 Further Separation into + and – Light-Front Fields

There have been many studies of light-front (infinite momentum frame) physics in the past forty years.[12] Light-front coordinates *cannot* be obtained by a Lorentz transformation, or by a superluminal transformation, from a standard set of coordinate system variables even in a limiting sense. Instead they are a defined set of variables that have been used to develop quantum field theories that have been shown to be equivalent to quantum field theories based on conventional coordinates. In particular, light-front quantum field theories have been shown to yield fully Lorentz covariant S matrix elements that are the same as S matrix elements calculated in the conventional way.

Light-front variables can be defined by:

$$x^{\pm} = (x^0 \pm x^3)/\sqrt{2}$$

$$(3.55)$$

---

[12] L. Susskind, Phys. Rev. **165**, 1535 (1968); K. Bardakci and M. B. Halpern Phys. Rev. **176**, 1686 (1968), S. Weinberg, Phys. Rev. **150**, 1313 (1966); J. Kogut and D. Soper, Phys. Rev. **D1**, 2901 (1970); J. D. Bjorken, J. Kogut, and D. Soper, Phys. Rev. **D3**, 1382 (1971); R. A. Neville and F. Rohrlich, Nuov. Cim. **A1**, 625 (1971); F. Rohrlich, Acta Phys Austr. Suppl. **8**, 277 (1971); S-J Chang, R. Root, and T-M Yan, Phys. Rev. **D7**, 1133 (1973); S-J Chang, and T-M Yan, Phys. Rev. **D7**, 1147 (1973); T-M Yan, Phys. Rev. **D7**, 1761 (1973); T-M Yan, Phys. Rev. **D7**, 1780 (1973); C. Thorn, Phys. Rev. **D19**, 639 (1979); and references therein.

$$\partial/\partial x^{\pm} \equiv \partial^{\mp} \equiv (\partial/\partial x^0 \pm \partial/\partial x^3)/\sqrt{2}$$

with the "transverse" coordinate variables, x^1 and x^2, unchanged.

The inner product of two 4-vectors has the form

$$x \cdot y = x^+ y^- + y^+ x^- - x^1 y^1 - x^2 y^2 \tag{3.56}$$

and the light-front definition of Dirac matrices is:

$$\gamma^{\pm} = (\gamma^0 \pm \gamma^3)/\sqrt{2} \tag{3.57}$$

with transverse matrices γ^1 and γ^2 defined as usual. Note the useful identity:

$$\gamma^{\pm 2} = 0$$

We define "+" and "–" tachyon fields with the projection operators:

$$R^{\pm} = \tfrac{1}{2}(I \pm \gamma^0 \gamma^3) \tag{3.58}$$

They are:

Left-handed, $\pm$ light-front fields: $\qquad \psi_{TL}{}^{\pm} = R^{\pm} C^- \psi_T$

$$\tag{3.59}$$

Right-handed, $\pm$ light-front fields: $\qquad \psi_{TR}{}^{\pm} = R^{\pm} C^+ \psi_T$

Now if we transform to light-front variables and fields as above we obtain the light-front free tachyon Lagrangian:

$$
\begin{aligned}
\mathcal{L}_T = {}& 2^{\frac{1}{2}} \psi_{TL}{}^{++\dagger} i\partial^- \psi_{TL}{}^+ + 2^{\frac{1}{2}} \psi_{TL}{}^{-\dagger} i\partial^+ \psi_{TL}{}^- - \psi_{TL}{}^{++\dagger} \gamma^0 i\gamma^j \partial^j \psi_{TL}{}^- - \psi_{TL}{}^{-\dagger} \gamma^0 i\gamma^j \partial^j \psi_{TL}{}^+ - \\
& - 2^{\frac{1}{2}} \psi_{TR}{}^{++\dagger} i\partial^- \psi_{TR}{}^+ - 2^{\frac{1}{2}} \psi_{TR}{}^{-\dagger} i\partial^+ \psi_{TR}{}^- + \psi_{TR}{}^{++\dagger} \gamma^0 i\gamma^j \partial^j \psi_{TR}{}^- + \psi_{TR}{}^{-\dagger} \gamma^0 i\gamma^j \partial^j \psi_{TR}{}^+ - \\
& - im[\psi_{TR}{}^{++\dagger} \gamma^0 \psi_{TL}{}^- - \psi_{TL}{}^{++\dagger} \gamma^0 \psi_{TR}{}^- + \psi_{TR}{}^{-\dagger} \gamma^0 \psi_{TL}{}^+ - \psi_{TL}{}^{-\dagger} \gamma^0 \psi_{TR}{}^+]
\end{aligned}
\tag{3.60}
$$

with implied sums over $j = 1, 2$. In contrast to the light-front tachyon Lagrangian we note the corresponding light-front "normal" Dirac fermion Lagrangian is

$$
\begin{aligned}
\mathcal{L}_{Dirac} = {}& 2^{\frac{1}{2}} \psi_L{}^{++\dagger} i\partial^- \psi_L{}^+ + 2^{\frac{1}{2}} \psi_L{}^{-\dagger} i\partial^+ \psi_L{}^- - \psi_L{}^{++\dagger} \gamma^0 i\gamma^j \partial^j \psi_L{}^- - \psi_L{}^{-\dagger} \gamma^0 i\gamma^j \partial^j \psi_L{}^+ - \\
& - 2^{\frac{1}{2}} \psi_R{}^{++\dagger} i\partial^- \psi_R{}^+ + 2^{\frac{1}{2}} \psi_R{}^{-\dagger} i\partial^+ \psi_R{}^- - \psi_R{}^{++\dagger} \gamma^0 i\gamma^j \partial^j \psi_R{}^- - \psi_R{}^{-\dagger} \gamma^0 i\gamma^j \partial^j \psi_R{}^+ - \\
& - im[\psi_R{}^{++\dagger} \gamma^0 \psi_L{}^- + \psi_L{}^{++\dagger} \gamma^0 \psi_R{}^- + \psi_R{}^{-\dagger} \gamma^0 \psi_L{}^+ + \psi_L{}^{-\dagger} \gamma^0 \psi_R{}^+]
\end{aligned}
\tag{3.61}
$$

The difference in signs between these Lagrangians will turn out to be a crucial factor in the derivation of features of the Standard Model later.

Returning to the tachyon Lagrangian eq. 3.60 we obtain equations of motion through the standard variational techniques:

$$2^{\frac{1}{2}}i\partial^{-}\psi_{TL}^{+} - \gamma^{0}i\gamma^{j}\partial^{j}\psi_{TL}^{-} + im\gamma^{0}\psi_{TR}^{-} = 0 \qquad (3.62)$$

$$2^{\frac{1}{2}}i\partial^{-}\psi_{TR}^{+} - \gamma^{0}i\gamma^{j}\partial^{j}\psi_{TR}^{-} + im\gamma^{0}\psi_{TL}^{-} = 0$$

$$2^{\frac{1}{2}}i\partial^{+}\psi_{TL}^{-} - \gamma^{0}i\gamma^{j}\partial^{j}\psi_{TL}^{+} + im\gamma^{0}\psi_{TR}^{+} = 0$$

$$2^{\frac{1}{2}}i\partial^{+}\psi_{TR}^{-} - \gamma^{0}i\gamma^{j}\partial^{j}\psi_{TR}^{+} + im\gamma^{0}\psi_{TL}^{+} = 0$$

Eqs. 3.62 show that ψ_{TL}^{-} and ψ_{TR}^{-} are dependent fields that are functions of ψ_{TL}^{+} and ψ_{TR}^{+} on the light-front where x^{+} equals a constant. They can be expressed in an integral form as well. (The independent fields ψ_{TL}^{+} and ψ_{TR}^{+} play a fundamental role in tachyon theory and are used to define "in" and "out" tachyon states in perturbation theory.)

The conjugate momenta are

$$\pi_{TL}^{+} = \partial \mathcal{L}/\partial(\partial^{-}\psi_{TL}^{+}) = 2^{\frac{1}{2}}i\psi_{TL}^{+\dagger} \qquad (3.63)$$

$$\pi_{TL}^{-} = \partial \mathcal{L}/\partial(\partial^{-}\psi_{TL}^{-}) = 0$$

$$\pi_{TR}^{+} = \partial \mathcal{L}/\partial(\partial^{-}\psi_{TR}^{+}) = -2^{\frac{1}{2}}i\psi_{TR}^{+\dagger} \qquad (3.64)$$

$$\pi_{TR}^{-} = \partial \mathcal{L}/\partial(\partial^{-}\psi_{TR}^{-}) = 0$$

Quantization on surfaces of constant x^{+} (light-front surfaces) has been shown to support satisfactory formulations of Quantum Electrodynamics and other quantum field theories. Thus x^{+} plays the role of the "time" variable in light-front quantized theories. So we will define canonical equal x^{+} anti-commutation relations for spin ½ tachyons.

The resulting canonical equal-light-front ($x^{+} = y^{+}$) anti-commutation relations of the independent fields are:

$$\{\psi_{TL}{}^{+\dagger}{}_{a}(x),\ \psi_{TL}{}^{+}{}_{b}(y)\} = 2^{-1}[C^{-}R^{+}]_{ab}\,\delta(x^{-} - y^{-})\delta^{2}(x - y) \qquad (3.65)$$

$$\{\psi_{TR}{}^{+\dagger}{}_{a}(x),\ \psi_{TR}{}^{+}{}_{b}(y)\} = -2^{-1}[C^{+}R^{+}]_{ab}\,\delta(x^{-} - y^{-})\delta^{2}(x - y) \qquad (3.66)$$

$$\{\psi_{TL}{}^{+}{}_{a}^{\dagger}(x),\ \psi_{TR}{}^{+}{}_{b}(y)\} = \{\psi_{TR}{}^{+}{}_{a}^{\dagger}(x),\ \psi_{TL}{}^{+}{}_{b}(y)\} = 0 \qquad (3.67)$$

$$\{\psi_{TL}{}^{+}{}_{a}(x),\ \psi_{TR}{}^{+}{}_{b}(y)\} = \{\psi_{TR}{}^{+}{}_{a}^{\dagger}(x),\ \psi_{TL}{}^{+\dagger}{}_{b}(y)\} = 0 \qquad (3.68)$$

where the factors of 2^{-1} are the result of the $2^{\frac{1}{2}}$ factor in eqs. 3.63 and 3.64, and the factor of $2^{-\frac{1}{2}}$ in the definition of x^{-} above.

If we compare eqs. 3.65 and 3.66 with the corresponding anti-commutation relations of *conventional Dirac* quantum fields:

$$\{\psi_{L}{}^{+\dagger}{}_{a}(x),\ \psi_{L}{}^{+}{}_{b}(y)\} = 2^{-1}[C^{-}R^{+}]_{ab}\,\delta(x^{-} - y^{-})\delta^{2}(x - y) \qquad (3.69)$$

$$\{\psi_R^{++\dagger}{}_a(x), \psi_R^{+}{}_b(y)\} = 2^{-1}[C^+R^+]_{ab}\,\delta(x^- - y^-)\delta^2(x - y) \qquad (3.70)$$

we see that the right-handed tachyon anti-commutation relation has a minus sign relative to the corresponding right-handed conventional anti-commutation relation. The right-handed tachyon anti-commutation relation with its minus sign will require compensating minus signs in its creation and annihilation Fourier component operators' anti-commutation relations.

The sign differences between the Lagrangian terms in eqs. 3.63 and 3.64 ultimately lead to parity violating features in the Standard Model Lagrangian and thus resolve the long-standing question:

3.4.4.3 Left-Handed Tachyons

The free, "+" light-front, left-handed tachyon wave function Fourier expansion is:

$$\psi_{TL}^{+}(x) = \sum_{\pm s}\int d^2p\,dp^+ N_{TL}^{+}(p)\theta(p^+)[b_{TL}^{+}(p, s)u_{TL}^{+}(p, s)e^{-ip\cdot x} + d_{TL}^{++}(p, s)v_{TL}^{+}(p, s)e^{+ip\cdot x}]$$

$$(3.71)$$

and its Hermitean conjugate is

$$\psi_{TL}^{++}(x) = \sum_{\pm s}\int d^2p\,dp^+ N_{TL}^{+}(p)\theta(p^+)\,[b_{TL}^{++}(p, s)u_{TL}^{++}(p,s)e^{+ip\cdot x} + d_{TL}^{+}(p, s)v_{TL}^{++}(p, s)e^{-ip\cdot x}]$$

$$(3.72)$$

where † indicates Hermitean conjugate, where

$$N_{TL}^{+}(p) = [2m|\mathbf{p}|/((2\pi)^3(p^+(p^+ - p^-) + p_\perp^2))]^{\frac{1}{2}} \qquad (3.73)$$

where the anti-commutation relations of the Fourier coefficient operators are

$$\begin{aligned}
\{b_{TL}^{+}(q,s), b_{TL}^{++}(p,s')\} &= \delta_{ss'}\delta^2(\mathbf{q} - \mathbf{p})\delta(q^+ - p^+)\\
\{d_{TL}^{+}(q,s), d_{TL}^{++}(p,s')\} &= \delta_{ss'}\delta^2(\mathbf{q} - \mathbf{p})\delta(q^+ - p^+)\\
\{b_{TL}^{+}(q,s), b_{TL}^{+}(p,s')\} &= \{d_{TL}^{+}(q,s), d_{TL}^{+}(p,s')\} = 0\\
\{b_{TL}^{++}(q,s), b_{TL}^{++}(p,s')\} &= \{d_{TL}^{++}(q,s), d_{TL}^{++}(p,s')\} = 0\\
\{b_{TL}^{+}(q,s), d_{TL}^{++}(p,s')\} &= \{d_{TL}^{+}(q,s), b_{TL}^{++}(p,s')\} = 0\\
\{b_{TL}^{++}(q,s), d_{TL}^{++}(p,s')\} &= \{d_{TL}^{+}(q,s), b_{TL}^{+}(p,s')\} = 0
\end{aligned} \qquad (3.74)$$

and where the spinors are

$$\begin{aligned}
u_{TL}^{+}(p, s) &= C^- R^+ S_L(\Lambda_L(\mathbf{p}))w^1(0)\\
u_{TL}^{+}(p, -s) &= C^- R^+ S_L(\Lambda_L(\mathbf{p}))w^2(0)\\
v_{TL}^{+}(p, s) &= C^- R^+ S_L(\Lambda_L(\mathbf{p}))w^3(0)
\end{aligned}$$

$$v_{TL}^{+}(p, -s) = C^{-} R^{+} S_{L}(\Lambda_{L}(\mathbf{p}))w^{4}(0) \tag{3.75}$$
$$u_{TL}^{++}(p, s) = w^{1T}(0)S_{L}^{\dagger}(\Lambda_{L}(\mathbf{p}))R^{+}C^{-}$$
$$u_{TL}^{++}(p, -s) = w^{2T}(0)S_{L}^{\dagger}(\Lambda_{L}(\mathbf{p}))R^{+}C^{-}$$
$$v_{TL}^{++}(p, s) = w^{3T}(0)S_{L}^{\dagger}(\Lambda_{L}(\mathbf{p}))R^{+}C^{-}$$
$$v_{TL}^{++}(p, -s) = w^{4T}(0)S_{L}^{\dagger}(\Lambda_{L}(\mathbf{p}))R^{+}C^{-}$$

where the superscript "T" indicates the transpose. (These spinors are described in Appendix 3-A.)

The canonical left-handed, light-front anti-commutation relation results in:

$$\{\psi_{TL}^{+}{}_{a}(x), \psi_{TL}^{+\dagger}{}_{b}(y)\} = \sum_{\pm s,s'} \int d^{2}pdp^{+}\int d^{2}p'dp'^{+} \, N_{TL}^{+}(p)N_{TL}^{+}(p')\theta(p^{+})\theta(p'^{+}) \cdot$$
$$\cdot [\{b_{TL}^{++}(p',s'),b_{TL}^{+}(p,s)\}u_{TL}^{+}{}_{a}(p,s)u_{TL}^{++}{}_{b}(p',s')e^{+ip'\cdot y - ip\cdot x} +$$
$$+ \{d_{TL}^{+}(p',s'),d_{TL}^{++}(p,s)\}v_{TL}^{+}{}_{a}(p,s)v_{TL}^{++}{}_{b}(p',s')e^{-ip'\cdot y + ip\cdot x}]$$

$$= \sum_{\pm s} \int d^{2}pdp^{+} N_{TL}^{+2}(p)\theta(p^{+})[u_{TL}^{+}{}_{a}(p,s)u_{TL}^{+}{}^{\dagger}{}_{b}(p,s)e^{+ip\cdot(y-x)} +$$
$$+ v_{TL}^{+}{}_{a}(p,s)v_{TL}^{++}{}_{b}(p,s)e^{-ip\cdot(y-x)}]$$

$$= -i\int d^{2}pdp^{+}\theta(p^{+})N_{TL}^{+2}(p)(2m|\mathbf{p}|)^{-1}\{[\,C^{-}R^{+}(i\not{p} - m)\gamma\cdot\mathbf{p}R^{+}C^{-}]_{ab}e^{+ip\cdot(y-x)} +$$
$$+ [C^{-}R^{+}(i\not{p} + m)\gamma\cdot\mathbf{p}R^{+}C^{-}]_{ab}e^{-ip\cdot(y-x)}\}$$

$$= -i\int d^{2}p_{\perp}\int_{0}^{\infty} dp^{+}N_{TL}^{+2}(p)\{[C^{-}R^{+}(ip^{+}(p^{+} - p^{-}) + ip_{\perp}^{2} - mp_{\perp}\cdot\gamma_{\perp})C^{-}]_{ab}e^{+ip^{+}(y^{-} - x^{-}) - ip_{\perp}\cdot(y_{\perp} - x_{\perp})} -$$
$$- [C^{-}R^{+}(-ip^{+}(p^{+} - p^{-}) - ip_{\perp}^{2} - mp_{\perp}\cdot\gamma_{\perp})C^{-}]_{ab}e^{-ip^{+}(y^{-} - x^{-}) + ip_{\perp}\cdot(y_{\perp} - x_{\perp})}\}/(2m|\mathbf{p}|)$$

$$= \int d^{2}p_{\perp}\int_{-\infty}^{\infty} dp^{+}N_{TL}^{+2}(p)[C^{-}R^{+}(p^{+}(p^{+} - p^{-}) + p_{\perp}^{2})]_{ab} \, e^{+ip^{+}(y^{-} - x^{-}) - ip_{\perp}\cdot(y_{\perp} - x_{\perp})}/(2m|\mathbf{p}|)$$

upon letting $p^{+} \to -p^{+}$ and $\mathbf{p}_{\perp} \to -\mathbf{p}_{\perp}$ in the second term after using $N_{TL}^{+2}(p)(p^{+}(p^{+} - p^{-}) + p_{\perp}^{2}) = 1$. The result

$$= \tfrac{1}{2}\int d^{2}p_{\perp}\int_{-\infty}^{\infty} dp^{+}(2\pi)^{-3}[C^{-}R^{+}]_{ab}e^{+ip^{+}(y^{-} - x^{-}) - ip_{\perp}\cdot(y_{\perp} - x_{\perp})}$$

$$= 2^{-1}[C^{-}R^{+}]_{ab}\delta(y^{-} - x^{-})\delta^{2}(\mathbf{y} - \mathbf{x}) \tag{3.76}$$

Therefore we obtain left-handed, light-front quantized tachyons with canonical commutation relations and localized tachyons. As a result we have a canonical Tachyon Quantum Field Theory unlike previous efforts.

3.4.4.4 Right-Handed Tachyons

The case of right-handed tachyons is similar to the left-handed case with only two differences: a minus sign in the creation and annihilation operator anti-commutation relations, and the use of right-handed projection operators. The right-handed tachyon wave function light-front Fourier expansion is:

$$\psi_{TR}^{+}(x) = \sum_{\pm s} \int d^2p\, dp^+ N_{TR}^{+}(p)\theta(p^+)[b_{TR}^{+}(p,\, s)u_{TR}^{+}(p,\, s)e^{-ip\cdot x} + d_{TR}^{++}(p,\, s)v_{TR}^{+}(p,\, s)e^{+ip\cdot x}]$$

$$(3.77)$$

and its Hermitean conjugate is

$$\psi_{TR}^{++}(x) = \sum_{\pm s} \int d^2p\, dp^+ N_{TR}^{+}(p)\theta(p^+)\, [b_{TR}^{++}(p,\, s)u_{TR}^{++}(p,\, s)e^{+ip\cdot x} + d_{TR}^{+}(p,\, s)v_{TR}^{++}(p,\, s)e^{-ip\cdot x}]$$

$$(3.78)$$

where $N_{TR}^{+}(p) = N_{TL}^{+}(p)$, where the anti-commutation relations of the Fourier coefficient operators are

$$\{b_{TR}^{+}(q,s),\, b_{TR}^{++}(p,s')\} = -\delta_{ss'}\delta^2(\mathbf{q} - \mathbf{p})\delta(q^+ - p^+) \qquad (3.79)$$
$$\{d_{TR}^{+}(q,s),\, d_{TR}^{++}(p,s')\} = -\delta_{ss'}\delta^2(\mathbf{q} - \mathbf{p})\delta(q^+ - p^+)$$
$$\{b_{TR}^{+}(q,s),\, b_{TR}^{+}(p,s')\} = \{d_{TR}^{+}(q,s),\, d_{TR}^{+}(p,s')\} = 0$$
$$\{b_{TR}^{++}(q,s),\, b_{TR}^{++}(p,s')\} = \{d_{TR}^{++}(q,s),\, d_{TR}^{++}(p,s')\} = 0$$
$$\{b_{TR}^{+}(q,s),\, d_{TR}^{++}(p,s')\} = \{d_{TR}^{+}(q,s),\, b_{TR}^{++}(p,s')\} = 0$$
$$\{b_{TR}^{++}(q,s),\, d_{TR}^{++}(p,s')\} = \{d_{TR}^{+}(q,s),\, b_{TR}^{+}(p,s')\} = 0$$

and where the spinors are

$$u_{TR}^{+}(p,\, s) = C^+R^+u_T(p,s) \qquad (3.80)$$
$$v_{TR}^{+}(p,\, s) = C^+R^+v_T(p,s) \qquad (3.81)$$

by Appendix 3-A (eq. 3-A.7).

The right-handed anti-commutation relation with the minus sign follows in particular because of the minus signs found earlier.

3.4.5 Interpretation of Tachyon Creation and Annihilation Operators

To properly discuss the physical interpretation of tachyon creation and annihilation operators we must first determine the Hamiltonian and momentum operators in terms of creation and annihilation operators.

The energy-momentum tensor density is the symmetrized version of

$$\mathcal{T}^{\mu\nu} = \sum_{i} \partial\mathcal{L}/\partial(\partial\chi_i/\partial x_\mu)\; \partial\chi_i/\partial x_\nu - g^{\mu\nu}\mathcal{L} \qquad (3.82)$$

where the sum over i is over the fields. The light-front Hamiltonian is

$$H \equiv P^- = T^{+-} = \int dx^- d^2x\, \mathcal{J}^{+-} \tag{3.83}$$

and the "momenta" are

$$P^+ = T^{++} = \int dx^- d^2x\, \mathcal{J}^{++} \tag{3.84}$$

$$P^i = T^{+i} = \int dx^- d^2x\, \mathcal{J}^{+i} \tag{3.85}$$

for i = 1,2.

The light-front, left-handed and right-handed tachyon Lagrangian $\mathcal{L}_T$ and its equations of motion imply

$$H = i2^{-\frac{1}{2}}\int dx^- d^2x\, [\psi_{TL}^{+\dagger}\partial^-\psi_{TL}^+ - \partial^-\psi_{TL}^{+\dagger}\psi_{TL}^+ + \psi_{TL}^{-\dagger}\partial^+\psi_{TL}^- - \partial^+\psi_{TL}^{-\dagger}\psi_{TL}^- -$$
$$- \psi_{TR}^{+\dagger}\partial^-\psi_{TR}^+ + \partial^-\psi_{TR}^{+\dagger}\psi_{TR}^+ - \psi_{TR}^{-\dagger}\partial^+\psi_{TR}^- + \partial^+\psi_{TR}^{-\dagger}\psi_{TR}^- + \text{mass terms}] \tag{3.86}$$

After substituting for the various fields we find the *independent fields* (which create the in and out particle states) have the Hamiltonian terms:

$$H = \sum_{\pm s}\int d^2p dp^+\, p^-[b_{TL}^{+\dagger}(p,s)b_{TL}^+(p,s) - d_{TL}^+(p,s)d_{TL}^{+\dagger}(p,s) - b_{TR}^{+\dagger}(p,s)b_{TR}^+(p,s) +$$
$$+ d_{TR}^+(p,s)d_{TR}^{+\dagger}(p,s)] \tag{3.87}$$

$$= \sum_{\pm s}\int d^2p dp^+\, p^-[b_{TL}^{+\dagger}(p,s)b_{TL}^+(p,s) + d_{TL}^{+\dagger}(p,s)d_{TL}^+(p,s) - b_{TR}^{+\dagger}(p,s)b_{TR}^+(p,s) -$$
$$- d_{TR}^{+\dagger}(p,s)d_{TR}^+(p,s)] \tag{3.88}$$

up to the usual infinite constants due to left-handed operator rearrangement and right-handed operator rearrangement that are discarded. Eq. 3.88 is the basis for our particle interpretation of tachyon creation and annihilation operators based on Dirac's hole theory. Dirac hole theory as applied in light-front coordinates assumes all negative p^- ("energy") states are filled.

3.4.5.1 Left-Handed Tachyon Creation and Annihilation Operators

1. We identify $b_{TL}^{+\dagger}(p,s)$ and $d_{TL}^+(p,s)$ as creation operators for left-handed tachyons. $b_{TL}^{+\dagger}(p,s)$ creates a positive p^- ("energy") state and $d_{TL}^+(p,s)$ creates a negative p^- ("energy") state.

2. $b_{TL}{}^{+}(p,s)$ and $d_{TL}{}^{++}(p,s)$ are the corresponding annihilation operators for left-handed tachyons. $b_{TL}{}^{+}(p,s)$ annihilates a positive p^{-} ("energy") state and $d_{TL}{}^{++}(p,s)$ annihilates a negative p^{-} ("energy") state.

3. We assume Dirac hole theory holds for the left-handed tachyon vacuum with all negative energy states filled. There is no tachyon energy gap as there is for Dirac fermions. There is also the problem that the left-handed tachyon vacuum is not invariant under ordinary Lorentz transformations or Superluminal transformations. *However if we confine ourselves to light-front coordinates for computations no ambiguity can result and the Lorentz covariant quantities that we calculate, such as the S matrix, are well-defined.*

4. Using tachyon hole theory we identify $b_{TL}{}^{+}(p,s)$ and $d_{TL}{}^{++}(p,s)$ as annihilation operators for left-handed tachyons. $b_{TL}{}^{+}(p,s)$ annihilates a positive p^{-} ("energy") state and $d_{TL}{}^{++}(p,s)$ annihilates a negative p^{-} ("energy") state – thus creating a hole in the tachyon sea that we view as the creation of a positive p^{-} ("energy"), left-handed antitachyon. $d_{TL}{}^{+}(p,s)$ annihilates a positive p^{-} ("energy"), left-handed anti-tachyon.

3.4.5.2 Right-Handed Tachyon Creation and Annihilation Operators

The anti-commutation relations of right-handed tachyon creation and annihilation operators and the right-handed Hamiltonian terms have the "wrong" sign compared to corresponding Dirac operators and left-handed tachyon operators. This situation is completely analogous to the situation of time-like photons in the covariant formulation of quantum Electrodynamics.[13] In the case of time-like photons it was possible to introduce an indefinite metric (Gupta-Bleuler formulation), and then to use the subsidiary condition $\partial A^{v}/\partial x^{v} = 0$ to reduce the dynamics of QED to the transverse components. Thus the time-like photons were intermediate artifacts needed to have a manifestly covariant formulation while QED observables depended solely on the transverse components of the electromagnetic field.

In the present case of free tachyons, and in leptonic ElectroWeak Theory there is no evident "subsidiary condition" to eliminate the right-handed tachyon fields. But since the only manner in which the right-handed leptonic tachyon fields[14] interact is through mass terms, which can be easily 'integrated out", right-handed leptonic tachyon fields are removed from the observable part of the leptonic ElectroWeak Theory by their "lack of interaction" with left-handed fields.

In the case of quark ElectroWeak Theory right-handed tachyon quark fields have charge $(-1/3)$ and thus experience an electromagnetic interaction as well as a Z

---

[13] Bogoliubov (1959) pp. 130-136.

[14] The tachyon fields are provisionally assumed to be neutrino fields in the leptonic sector, and d, s and b quarks in the quark sector.

interaction. However, since quarks are totally confined, right-handed tachyon quarks will not be able to continuously emit photons or Z's due to energy conservation and their confinement to bound states of fixed positive energy. Earlier, when we consider complex Lorentz group boosts, we will suggest that quarks may not consist of Dirac particles or tachyons of the type considered up to this point in this chapter. Rather they may be variants on Dirac particles and tachyons satisfying different dynamical equations. However, the preceding comments on quarks would still apply.

Thus right-handed tachyons are analogous to time-like photons – necessary theoretically but prevented from causing a negative energy disaster by the forms of their interactions. We discuss this subject in more detail in the following chapters [*where a much different approach is followed.*]

3.4.6 Tachyon Feynman Propagator

In this section we develop the light-front propagator for tachyons. We begin with a subsection describing the light-front propagators of Dirac fields.

3.4.6.1 Dirac Field Light-Front Propagators

The light-front Feynman propagator for the ψ^+ field of a Dirac fermion is

$$iS^+_F(x,y)\gamma^0 = \theta(x^+ - y^+)<0|\psi^+(x)\psi^{+\dagger}(y)|0> - \theta(y^+ - x^+)<0|\psi^{+\dagger}(y)\psi^+(x)|0> \quad (3.89)$$

and does not contain a non-covariant piece due to the projection operators:

$$iS^+_F(x,y) = \int d^2p dp^+ \theta(p^+)[1/(2(2\pi)^3 p^+)]\{\theta(x^+ - y^+)[R^+(\not{p} + m)R^-]\,e^{-ip\cdot(x-y)} +$$
$$+ \theta(y^+ - x^+)[R^+(-\not{p} + m)R^-]e^{+ip\cdot(x-y)}\}$$
$$= R^+ iS_F(x,y)R^- \quad (3.90)$$

where $S_F(x,y)$ is the usual Feynman propagator.

The light-front Feynman propagator for a *left-handed* <u>Dirac</u> field ψ^+ is
$$iS^+_{LF}(x,y) = \int d^2p dp^+ \theta(p^+)[1/(2(2\pi)^3 p^+)]\{\theta(x^+ - y^+)[C^-R^+(\not{p} + m)R^-C^-]e^{-ip\cdot(x-y)} +$$
$$+ \theta(y^+ - x^+)[C^-R^+(-\not{p} + m)R^-C^-]e^{+ip\cdot(x-y)}\}$$
$$= C^-R^+ iS_F(x,y)R^-C^- \quad (3.91)$$

3.4.6.2 Tachyon Field Light Front Propagators

Turning now to tachyons, the light-front Feynman propagator for the left-handed ψ_{TL}^+ *tachyon* field is (using the previous Fourier expansion of the left-handed tachyon field):

$$iS^+_{TLF}(x,y) = \theta(x^+ - y^+)<0|\psi_{TL}^+(x)\psi_{TL}^{+\dagger}(y)\gamma^0|0> - \theta(y^+ - x^+)<0|\psi_{TL}^{+\dagger}(y)\gamma^0\psi_{TL}^+(x)|0>$$
$$= -i\int d^2p dp^+ \theta(p^+)N_{TL}^{+2}(2m|\mathbf{p}|)^{-1}C^-R^+\{\theta(x^+ - y^+)[(i\not{p} - m)\gamma\cdot\mathbf{p}]e^{-ip\cdot(x-y)} +$$
$$+ \theta(y^+ - x^+)[(i\not{p} + m)\gamma\cdot\mathbf{p}]e^{+ip\cdot(x-y)}\}R^+C^-\gamma^0$$

If we define the on-shell momentum variable

$$p_0^- = (p_0^1 p_0^1 + p_0^2 p_0^2 - m^2)/(2p_0^+), \quad p_0^+ = p^+, \quad p_0^j = p^j \text{ (for } j = 1, 2), \quad p_{\perp 0}^2 = p_0^j p_0^j$$

and

$$\not{p}_0 = p_0 \cdot \gamma$$

then the above equation can be rewritten as

$$S^+{}_{TLF}(x,y) = -C^-R^+\!\int d^4p[32\pi^4(p_0^+(p_0^+ - p_0^-) + p_{0\perp}^2)]^{-1}e^{-ip\cdot(x-y)}\{\theta(p^+)(i\not{p}_0 - m)\gamma\cdot\mathbf{p}_0]/[p^- -$$
$$- p_0^- + i\varepsilon] + \theta(-p^+)(i\not{p}_0 + m)\gamma\cdot\mathbf{p}_0]/[p^- + p_0^- - i\varepsilon]\}R^+C^-\gamma^0$$
$$= -\tfrac{1}{2}\,i\!\int d^4p(2\pi)^{-4}[C^-R^+(i\not{p} - m)\gamma\cdot\mathbf{p}R^+C^-\gamma^0]e^{-ip\cdot(x-y)}[(p^2 + m^2 + i\varepsilon)(p^+(p^+ - p^-) + p_\perp^2))]^{-1}$$

and using $C^-R^+(i\not{p} - m)\gamma\cdot\mathbf{p}R^+C^- = i\,C^-R^+(p^+(p^+ - p^-) + p_\perp^2)$ we find

$$iS^+{}_{TLF}(x,y) = \tfrac{1}{2}C^-R^+\gamma^0\!\int d^4p(2\pi)^{-4}\,p^+e^{-ip\cdot(x-y)}/(p^2 + m^2 + i\varepsilon) \tag{3.92}$$

Similarly the light-front Feynman propagator for the right-handed ψ_{TR}^+ tachyon field is

$$iS^+{}_{TRF}(x,y) = \theta(x^+ - y^+)\langle 0|\psi_{TR}^+(x)\psi_{TR}^{++}(y)\gamma^0|0\rangle - \theta(y^+ - x^+)\langle 0|\psi_{TR}^{++}(y)\gamma^0\psi_{TR}^+(x)|0\rangle$$
$$= -\tfrac{1}{2}C^+R^+\gamma^0\!\int d^4p(2\pi)^{-4}\,p^+e^{-ip\cdot(x-y)}/(p^2 + m^2 + i\varepsilon) \tag{3.93}$$

where the relative minus sign between eqs. 3.92 and 3.93 is due to the relative minus signs of the Fouier component operator anti-commutation relations.

Thus we find *tachyon* pole terms in the tachyon propagators as one would expect.

3.5 Complex Space and 3-Momentum & Real-Valued Energy Fermions (Quarks)

In this section we will use L_C boosts to develop a wider set of dynamical equations for free spin ½ fermions with real-valued energy and complex-valued 3-momentum.[15] Earlier we defined L_C boosts with

$$\Lambda_C(\mathbf{v}_c) = \exp[i\omega\hat{\mathbf{w}}\cdot\mathbf{K}] \tag{3.94}$$
$$\omega = (\omega_r^2 - \omega_i^2 + 2i\omega_r\omega_i\,\hat{\mathbf{u}}_r\cdot\hat{\mathbf{u}}_i)^{\tfrac{1}{2}} \tag{3.95}$$
$$\hat{\mathbf{w}} = (\omega_r\hat{\mathbf{u}}_r + i\omega_i\hat{\mathbf{u}}_i)/\omega \tag{3.96}$$
$$\hat{\mathbf{w}}\cdot\hat{\mathbf{w}} = \hat{\mathbf{u}}_r\cdot\hat{\mathbf{u}}_r = \hat{\mathbf{u}}_i\cdot\hat{\mathbf{u}}_i = 1 \tag{3.97}$$

---

[15] The complexon theory that we develop and use for quark dynamics in the Standard Model is <u>not</u> required. Our Standard Model could use Dirac fermion dynamics for the up-type quarks and tachyon dynamics for down-type quarks. We choose to use complexon dynamics for all quark types because they have an internal SU(3)-like structure suggestive of color SU(3). More importantly, their spin dynamics is different and thus may resolve the differences between theory and experiment – particularly for the deep inelastic parton spin-dependent structure functions. <u>This section also changes significantly later.</u>

$$v_c = \hat{w}\tanh(\omega) \tag{3.98}$$

3.5.1 L_C Spinor "Normal" Lorentz Boosts & More Spin ½ Particle Types

Spinor boost transformations were used in previous sections to develop the dynamical equations for Dirac fields and tachyon fields. In this section we will use L_C spinor boosts to generate additional fermion field dynamical equations.

The form of the L_C spinor boost transformation corresponding to the coordinate transformation is:

$$S_C(\omega, v_c) = \exp(-i\omega\sigma_{0k}\hat{w}_k/2) = \exp(-\omega\gamma^0\boldsymbol{\gamma}\cdot\hat{w}/2)$$
$$= \cosh(\omega/2)I + \sinh(\omega/2)\gamma^0\boldsymbol{\gamma}\cdot\hat{w} \tag{3.99}$$

The inverse transformation is

$$S_C^{-1}(\omega, v_c) = \gamma^2\gamma^0 K^{-1}S_C^\dagger K\gamma^0\gamma^2 = \gamma^2\gamma^0 S_C^{\,T}\gamma^0\gamma^2 = \exp(\omega\gamma^0\boldsymbol{\gamma}\cdot\hat{w}/2)$$
$$= \cosh(\omega/2)I - \sinh(\omega/2)\gamma^0\boldsymbol{\gamma}\cdot\hat{w} \tag{3.100}$$

where the superscript T denotes the transpose and K is the complex conjugation operator (that also appears in the time-reversal operator). Note that S_C is not unitary just as in previous cases considered in this chapter.

We now redo the development of spin ½ dynamical equations of motion of earlier sections for this more general case of complex ω and $\hat{w}$. Again we apply a boost to a Dirac equation for a positive energy plane wave particle of mass m at rest:

$$0 = S_C(\omega, v_c))(m\gamma^0 - m)e^{-imt}w(0)$$
$$= [mS_C\gamma^0 S_C^{-1} - m]e^{-imt}S_C w(0) \tag{3.101}$$

where $S_C = S_C(\omega, \hat{w})$. After some algebra

$$mS_C\gamma^0 S_C^{-1} = m[\cosh(\omega)\gamma^0 - \sinh(\omega)\boldsymbol{\gamma}\cdot\hat{w}] \tag{3.102}$$

3.5.1.1 Case 1: Parallel Real and Imaginary Relative Vectors

If the real and imaginary relative vectors parts of $\hat{w}$, namely $\hat{u}_r$ and $\hat{u}_i$, are parallel, then $\hat{u}_r\cdot\hat{u}_i = 1$ and

$$\omega = \omega_r + i\omega_i \tag{3.103}$$

Eq. 3.102 can be re-expressed as

$$mS_C\gamma^0 S_C^{-1} = m[\cosh(\omega_r)\cos(\omega_i) + i\sinh(\omega_r)\sin(\omega_i)]\gamma^0 - m[\sinh(\omega_r)\cos(\omega_i) + i\cosh(\omega_r)\sin(\omega_i)]\boldsymbol{\gamma}\cdot\hat{u}_r \tag{3.104}$$

or equivalently

$$mS_C\gamma^0 S_C^{-1} = \cos(\omega_i)\boldsymbol{\gamma}\cdot p_r + i\sin(\omega_i)\boldsymbol{\gamma}\cdot p_i \tag{3.105}$$

where

$$p_r^{\,0} = m\cosh(\omega_r) \qquad\qquad p_i^{\,0} = m\sinh(\omega_r) \tag{3.106}$$

and

$$\mathbf{p_r} = m\hat{\mathbf{u}}_r \sinh(\omega_r) \qquad\qquad \mathbf{p_i} = m\hat{\mathbf{u}}_r \cosh(\omega_r) \qquad (3.107)$$

If $\omega_i = 0$, then we recover the momentum space Dirac equation. If $\omega_i = \pi/2$, then we obtain the left-handed momentum space tachyon equation. Since the range of ω_i is [0, ∞> (due to the cut along the real ω-plane axis) eq. 3.105 corresponds to the results of the Left-Handed Lorentz boost part discussed earlier.

3.5.1.2 Case 2: Anti-Parallel Real and Imaginary Relative Vectors

  If the real and imaginary relative vectors parts of $\hat{\mathbf{w}}$, $\hat{\mathbf{u}}_r$ and $\hat{\mathbf{u}}_i$, are anti-parallel $\hat{\mathbf{u}}_r = -\hat{\mathbf{u}}_i$, then $\hat{\mathbf{u}}_r \cdot \hat{\mathbf{u}}_i = -1$ and

$$\omega = \omega_r - i\omega_i \qquad (3.108)$$

We can then express eq. 3.105 as

$$mS_C\gamma^0 S_C^{-1} = m[\cosh(\omega_r)\cos(\omega_i) - i\sinh(\omega_r)\sin(\omega_i)]\gamma^0 - m[\sinh(\omega_r)\cos(\omega_i) -$$
$$- i\cosh(\omega_r)\sin(\omega_i)]\gamma \cdot \hat{\mathbf{u}}_r \qquad (3.109)$$

or

$$mS_C\gamma^0 S_C^{-1} = \cos(\omega_i)\gamma \cdot \mathbf{p_r} - i\sin(\omega_i)\gamma \cdot \mathbf{p_i} \qquad (3.110)$$

where

$$p_r^{\,0} = m\cosh(\omega_r) \qquad\qquad p_i^{\,0} = m\sinh(\omega_r) \qquad (3.111)$$

and

$$\mathbf{p_r} = m\hat{\mathbf{u}}_r \sinh(\omega_r) \qquad\qquad \mathbf{p_i} = m\hat{\mathbf{u}}_r \cosh(\omega_r) \qquad (3.112)$$

If $\omega_i = 0$, then we again recover the momentum space Dirac equation, If $\omega_i = \pi/2$, then we obtain the right-handed momentum space tachyon equation. (The range of ω_i is again [0, ∞>.)

Note: Since the matrix elements in the boost depend on $\gamma = (1 - \beta^2)^{-\frac{1}{2}}$ with a singularities at $\beta = \pm1$, which in turn corresponds to $\omega = \pm\infty$, there is a branch cut along the ω axis in the complex ω-plane. Therefore we point out again the product of three Left-handed transformations is not equivalent to a Right-handed transformation.

3.5.1.3 Case 3: Complexons: A New Type of Particle with Perpendicular Real and Imaginary 3-Momenta

  If the real and imaginary relative vectors parts of $\hat{\mathbf{w}}$, namely $\hat{\mathbf{u}}_r$ and $\hat{\mathbf{u}}_i$, are perpendicular, $\hat{\mathbf{u}}_r \cdot \hat{\mathbf{u}}_i = 0$, then

$$\omega = (\omega_r^2 - \omega_i^2)^{\frac{1}{2}} \qquad (3.113)$$

Thus ω is either pure real ($\omega_r \geq \omega_i$) or pure imaginary ($\omega_r < \omega_i$).

  The momentum space equation generated by the corresponding L_C spinor boost is

$$\{m\cosh(\omega)\gamma^0 - m\sinh(\omega)\gamma \cdot (\omega_r\hat{\mathbf{u}}_r + i\omega_i\hat{\mathbf{u}}_i)/\omega - m\}e^{-ip\cdot x}w_c(p) = 0 \qquad (3.114)$$

Defining the momentum 4-vector

$$p = (p^0, \mathbf{p}) \tag{3.115}$$

where

$$p^0 = m\cosh(\omega) \qquad\qquad \mathbf{p} = \mathbf{p_r} + i\mathbf{p_i} \tag{3.116}$$

$$\mathbf{p_r} = m\omega_r\hat{\mathbf{u}}_r\sinh(\omega)/\omega \qquad \mathbf{p_i} = m\omega_i\hat{\mathbf{u}}_i\sinh(\omega)/\omega \tag{3.117}$$

and

$$\mathbf{p_r}\cdot\mathbf{p_i} = 0 \tag{3.118}$$

then we obtain a positive energy Dirac-like equation with complex 3-momentum

$$[\mathbf{p}\cdot\gamma - m]e^{-ip\cdot x}w_c(p) = 0$$

or, explicitly,
$$\tag{3.119}$$

$$[p^0\gamma^0 - (\mathbf{p_r} + i\mathbf{p_i})\cdot\gamma - m]e^{-ip\cdot x}w_c(p) = 0$$

with a complex 3-momentum $\mathbf{p}$ and the 4-momentum mass shell condition:

$$p^2 = p^{0\,2} - \mathbf{p_r}\cdot\mathbf{p_r} + \mathbf{p_i}\cdot\mathbf{p_i} = m^2 \tag{3.120}$$

Note

$$|\mathbf{v}| = |\mathbf{p}|/p^0 = [(\mathbf{p_r} + i\mathbf{p_i})\cdot(\mathbf{p_r} + i\mathbf{p_i})]^{\frac{1}{2}}/p^0 = \tanh(\omega) \tag{3.121}$$

and thus the Lorentz factor

$$\gamma = \cosh(\omega) \tag{3.122}$$

Eq. 3.119 is the momentum space equivalent of the wave equation

$$[i\gamma^0\partial/\partial t + i\gamma\cdot(\nabla_r + i\nabla_i) - m]\psi_C(t, \mathbf{x_r}, \mathbf{x_i}) = 0 \tag{3.123}$$

where

$$x_c = (t, \mathbf{x_r} - i\mathbf{x_i}) \tag{3.123a}$$

and where the grad operators ∇_r and ∇_i are with respect to $\mathbf{x_r}$ and $\mathbf{x_i}$ respectively. Since $\hat{\mathbf{u}}_r\cdot\hat{\mathbf{u}}_i = 0$, we see that there is a subsidiary condition on the wave function

$$\nabla_r\cdot\nabla_i\,\psi_C(t, \mathbf{x_r}, \mathbf{x_i}) = 0 \tag{3.124}$$

We will call the particles satisfying eqs.3.123 and 3.124 *complexons*. In addition eq. 3.118 implies the anti-commutation relation

$$\{\gamma\cdot\mathbf{p_r}, \gamma\cdot\mathbf{p_i}\} = 0 \tag{3.125}$$

which in turn implies

$$\gamma\cdot\nabla_r\gamma\cdot\nabla_i\psi_C(t, \mathbf{x_r}, \mathbf{x_i}) = \gamma\cdot\nabla_i\gamma\cdot\nabla_r\psi_C(t, \mathbf{x_r}, \mathbf{x_i}) = 0 \tag{3.126}$$

We note that eq. 3.125 is covariant under the real Lorentz group and eq. 3.126 can be easily put into covariant form since the difference of these 4-vectors squared is a real Lorentz group invariant: $[\gamma^0\partial/\partial t + \gamma\cdot(\nabla_r + i\nabla_i)]^2 - [\gamma^0\partial/\partial t + i\gamma\cdot(\nabla_r - i\nabla_i)]^2 = 4\nabla_r\cdot\nabla_i$.

Before considering a Lagrangian formulation and the Fourier operator representation of $\psi_C(t, \mathbf{x_r}, \mathbf{x_i})$ we will define the spinors and associated real and imaginary spin operators.

The spinor generated from a spin up Dirac spinor at rest by a complex boost is

$$w_c(p) = S_C(p)w(0) = [\cosh(\omega/2)I + \sinh(\omega/2)\gamma^0\gamma\cdot\hat{\mathbf{w}}]w(0) \qquad (3.127)$$

Following a procedure similar to Appendix 3-A (which the reader may wish to examine first) we define four spinors for Dirac particles at rest:

$$w^k(0) = \begin{bmatrix} \delta_{1k} \\ \delta_{2k} \\ \delta_{3k} \\ \delta_{4k} \end{bmatrix} \qquad (2\text{-A.2})$$

where Kronecker deltas appear in the brackets. Then, by applying eq. 3.127 to the spinors defined by eq. 3-A.2, we find the L_C spinors

$$S_C w^k(0) = w_{Cr}{}^k(p) + iw_{Ci}{}^k(p) \qquad (3.128)$$

where

$$\begin{aligned} S_{Cr} &= \cosh(\omega/2)I + (\omega_r/\omega)\sinh(\omega/2)\gamma^0\gamma\cdot\hat{\mathbf{u}}_r \\ &= [(m + E)/(2m)]^{\frac{1}{2}}I + [m(m + E)]^{-\frac{1}{2}}\gamma^0\gamma\cdot\mathbf{p}_r = aI + b\gamma^0\gamma\cdot\mathbf{p}_r \end{aligned} \qquad (3.129)$$

Thus the "real" spinors $w_{Cr}{}^k(p)$ are the columns of

$$S_{Cr} = \begin{array}{cccc} \underline{w_{Cr}{}^1(p)} & \underline{w_{Cr}{}^2(p)} & \underline{w_{Cr}{}^3(p)} & \underline{w_{Cr}{}^4(p)} \\ \begin{bmatrix} a & 0 & bp_{r\,z} & bp_{r-} \\ 0 & a & bp_{r+} & -bp_{r\,z} \\ bp_{r\,z} & bp_{r-} & a & 0 \\ bp_{r+} & -bp_{r\,z} & 0 & a \end{bmatrix} \end{array}$$

$$(3.130)$$

where $p_{r\pm} = p_{r\,x} \pm ip_{r\,y}$. The "imaginary" spinors are the columns of

$$S_{Ci} = (\omega_i/\omega)\sinh(\omega/2)\gamma^0\gamma\cdot\hat{\mathbf{u}}_i = [m(m + E)]^{-\frac{1}{2}}\gamma^0\gamma\cdot\mathbf{p}_i = b\gamma^0\gamma\cdot\mathbf{p}_i \qquad (3.131)$$

$$\underline{w_{Ci}{}^1(p)} \qquad \underline{w_{Ci}{}^2(p)} \qquad \underline{w_{Ci}{}^3(p)} \qquad \underline{w_{Ci}{}^4(p)}$$

$$S_{Ci} = \begin{bmatrix} 0 & 0 & bp_{iz} & bp_{i-} \\ 0 & 0 & bp_{i+} & -bp_{iz} \\ bp_{iz} & bp_{i-} & 0 & 0 \\ bp_{i+} & -bp_{iz} & 0 & 0 \end{bmatrix}$$

$$(3.132)$$

where $p_{i\pm} = p_{ix} \pm ip_{iy}$.

Eqs. 3.127 through 3.132 imply that the wave function solution of eq. 3.123, subject to the subsidiary condition eq. 3.124, is[16, 17]

$$\psi_C(x_r, x_i) = \sum_{\pm s} \int d^3p_r d^3p_i \, N_C(p)\delta(\mathbf{p_r}\cdot\mathbf{p_i}/m^2)[b_C(p,s)u_C(p, s)e^{-i(p\cdot x + p^*\cdot x^*)/2} +$$
$$+ d_C^\dagger(p,s)v_C(p, s)e^{+i(p\cdot x + p^*\cdot x^*)/2}]$$

$$(3.133)$$

where $\mathbf{p} = \mathbf{p_r} + i\mathbf{p_i}$ (eq. 3.95), $x = x_r - ix_i$, $p\cdot x = p^0 x^0 - \mathbf{p}\cdot\mathbf{x}$, and where we use

$$(p\cdot x + p^*\cdot x^*)/2 = p^0 x^0 - \mathbf{p_r}\cdot\mathbf{x_r} - \mathbf{p_i}\cdot\mathbf{x_i}$$

$$(3.134)$$

in the exponentials in order to avoid divergences that would appear in the calculation of the equal-time commutator, the Feynman propagator and other quantities of interest after second quantization. Note that

$$(\nabla_r + i\nabla_i)e^{-i(p\cdot x + p^*\cdot x^*)/2} = i(\mathbf{p_r} + i\mathbf{p_i})e^{-i(p\cdot x + p^*\cdot x^*)/2}$$

$$(3.135)$$

and

$$(\nabla_r + i\nabla_i)e^{-ip^*\cdot x^*} = 0$$

$$(3.136)$$

for all p.

The wave function's conjugate (the Hermitean conjugate modified by letting $x_i \rightarrow -x_i$ in addition to Hermitean conjugation) is

$$\psi_C^\dagger(x) = \psi_C^\dagger(x_r, -x_i) = \sum_{\pm s} \int d^3p_r d^3p_i \, \delta(\mathbf{p_r}\cdot\mathbf{p_i}/m^2)N_C(p^*)\cdot$$
$$\cdot[b_C^\dagger(p^*,s)u_C^\dagger(p^*,s)e^{+i(p\cdot x^* + p^*\cdot x)/2} + d_C(p^*,s)v_C^\dagger(p^*,s)e^{-i(p\cdot x^* + p^*\cdot x)/2}]$$

$$(3.137)$$

---

[16] Note that when $|\mathbf{p_i}| \geq |\mathbf{p_r}|$ (for imaginary $\omega = (\omega_r^2 - \omega_i^2)^{\frac{1}{2}}$) the 3-momentum becomes imaginary $\mathbf{p}\cdot\mathbf{p} < 0$. However, since we will be identifying confined quarks with this type of particle – much modified by a confining color quark interaction – the issue of an imaginary 3-momentum in the hypothetical free quark case becomes moot. We note the energy gap between positive and negative energy states disappears so $E = 0$ is possible. Thus real Lorentz transformations can mix positive and negative energy states. The solution is to do all calculations in the light-front frame as we do for tachyons. Then the mixing issue is resolved. In the present case we second quantize on the "time-front" for illustrative purposes.

[17] We scale $\mathbf{p_r}\cdot\mathbf{p_i}$ with m^2 in the delta function for convenience. All fermions have at least a minimal mass – the mass of the qube.

where $\mathbf{p} = \mathbf{p_r} + i\mathbf{p_i}$, $\mathbf{x} = \mathbf{x_r} - i\mathbf{x_i}$, $p \cdot x = p^0 x^0 - \mathbf{p} \cdot \mathbf{x}$, and † indicates Hermitean Hermitean conjugation.

The spinors are

$$u_C(p, s) = S_C(p)w^1(0)$$
$$u_C(p, -s) = S_C(p)w^2(0)$$
$$v_C(p, s) = S_C(p)w^3(0)$$
$$v_C(p, -s) = S_C(p)w^4(0) \tag{3.138}$$
$$u_C^\dagger(p^*, s) = w^{1T}(0)S_C^\dagger(p^*) = w^{1T}(0)S_C(p)$$
$$u_C^\dagger(p^*, -s) = w^{2T}(0)S_C^\dagger(p^*) = w^{2T}(0)S_C(p)$$
$$v_C^\dagger(p^*, s) = w^{3T}(0)S_C^\dagger(p^*) = w^{3T}(0)S_C(p)$$
$$v_C^\dagger(p^*, -s) = w^{4T}(0)S_C^\dagger(p^*) = w^{4T}(0)S_C(p)$$

with the superscript "T" indicating the transpose. Note that

$$S_C^\dagger(p^*) = [S_C(p^*)]^\dagger = S_C(p) \tag{3.139}$$

The normalization factor $N_C(p)$ is

$$N_C(p) = [2m/((2\pi)^6 p^0)]^{\frac{1}{2}} \tag{3.140}$$

Since $\mathbf{p_r} = \mathbf{p_i} = 0$ in the particle rest frame prior to the complex group boost, the boosted particle spin 4-vector s^μ satisfies

$$s^\mu p_r{}^\mu = s^\mu p_i{}^\mu = 0 \tag{3.141}$$

Note that s^μ is itself complex[18] and, if the spin points in the z-direction prior to the complex boost, then the boosted s^μ has the form

$$s^\mu = (-\sinh(\omega)\hat{w}_z, (0,0,1) + (\cosh(\omega) - 1)\hat{w}_z\hat{\mathbf{w}}) \tag{3.142}$$

with $\hat{\mathbf{w}}$ defined earlier: $\hat{\mathbf{w}} = (\omega_r\hat{\mathbf{u}}_r + i\omega_i\hat{\mathbf{u}}_i)/\omega = \mathbf{p}/(m\sinh(\omega))$.

3.5.1.4 A Global SU(3) Symmetry Revealed

Before proceeding to consider the second quantization of this case, we will consider a global SU(3) symmetry implicit in the previous equations. The defining property of the group SU(3) is that it preserves the invariance of inner products of complex 3-vectors of the form:

$$u^* \cdot v = u^1{}^* v^1 + u^2{}^* v^2 + u^3{}^* v^3 \tag{3.143}$$

---

[18] This feature of partons, which is not present in ordinary Dirac particles, might be the source of the discrepancies between theory and experiment in deep inelastic parton spin physics which is based on conventional real parton spins.

If we examine the dynamical equation eq. 3.123 we see that the differential operator is invariant under an SU(3) transformation U (using $\nabla_c = (\nabla_c{*})^* = \mathbf{D}_c{*}$)

$$[i\gamma^0\partial/\partial t + i\mathbf{D}_c{*}\cdot\gamma - m] = [i\gamma^0\partial/\partial t + i\mathbf{D}_c'{*}\cdot\gamma' - m] \tag{3.144}$$

where

$$\mathbf{D}_c{*} = \nabla_c = \nabla_r + i\nabla_i$$

and

$$\gamma'^a = U^{ab}\gamma'^b$$
$$\mathbf{D}_c'{*}^a = \mathbf{D}_c'{*}^b U^{\dagger ab}$$

where U is a global SU(3) transformation and $U^\dagger = U^{-1}$. By theorem[19] all 4×4 γ matrices such as γ' are equivalent up to a unitary transformation V. Thus $V^\dagger\gamma'V = \gamma$ and eq. 3.144 is equivalent to

$$[i\gamma^0\partial/\partial t + i\mathbf{D}_c{*}\cdot\gamma - m] = [i\gamma^0\partial/\partial t + i\mathbf{D}_c'{*}\cdot\gamma - m] \tag{3.145}$$

$$= [i\gamma^0\partial/\partial t + i\nabla_c'\cdot\gamma - m]$$

where $\nabla_c'_a = U^{ab}\nabla_{cb}$, This demonstrates that eq. 3.123 is invariant under an SU(3) transformation if

$$\psi_C(t, \mathbf{x}_c) = \psi_C(t, U\mathbf{x}_c) = \psi_C'(t, \mathbf{x}_c') \tag{3.146}$$

where $\psi_C(t, \mathbf{x}_c) \equiv \psi_C(t, \mathbf{x}_r, \mathbf{x}_i)$.

The subsidiary condition eq. 3.124 can be seen to transform as

$$\nabla_r\cdot\nabla_i\,\psi_C(t, \mathbf{x}_c) = \nabla_r{*}\cdot\nabla_i\,\psi_C(t, \mathbf{x}_c) = \nabla_r'{*}\cdot\nabla_i'\psi_C'(t, \mathbf{x}_c') = 0 \tag{3.147}$$

under an SU(3) rotation. The invariance of the orthogonality condition is preserved.

The wave function (eq. 3.123) transforms in the following way under the SU(3) transformation U. If we define

$$q{*}^\mu = (q^0, \mathbf{q}{*}) = (p^0, \mathbf{p}_r + i\mathbf{p}_i) = (p^0, \mathbf{p}) = p^\mu \tag{3.148}$$

then eq. 3.133 can be rewritten in an invariant form under a SU(3) transformation:

$$\psi_C(x) = \sum_{\pm s}\int d^3q_r d^3q_i\, N_C(p^0)\delta(\mathbf{q}_r{*}\cdot\mathbf{q}_i/m^2)[b_C(q{*},s)u_C(q{*},s)e^{-i(q{*}\cdot x + q\cdot x{*})/2} +$$
$$+ d_C^\dagger(q{*},s)v_C(q{*},s)e^{+i(q{*}\cdot x + q\cdot x{*})/2}] \tag{3.149}$$

---

[19] R. H. Good, Rev. Mod. Phys., **27**, 187 (1955).

where $x = x_c$ subject to an examination of the transformation properties of the Fourier coefficients and spinors. Note both terms in each exponential are separately invariant under global SU(3). (Note also $\mathbf{q}_r^* = \mathbf{q}_r$ since $\mathbf{q}_r$ is real.)

From the form of S_C above it is clear that an argument similar to that for the dynamical equations shows S_C is invariant under an SU(3) transformation and thus their spinors are also invariant under SU(3) transformations. The Fourier coefficients, if second quantized in a direct generalization of the usual manner, have covariant anti-commutation relations under an SU(3) transformation. For example

$$\{b_C(q,s),\, b_C^{\dagger}(q'^*,s')\} = \delta_{ss'}\delta^3(q_r - q'_{r'})\delta^3(q_i - q'_{i'}) \tag{3.150}$$

Under an SU(3) transformation, $z = Uq$ and $z' = Uq'$, the right side of eq. 3.150 transforms to

$$\delta^3(q_r - q'_{r'})\delta^3(q_i - q'_{i'}) \rightarrow \delta^3(z_r - z'_{r'})\delta^3(z_i - z'_{i'})/|\partial(q)/\partial(z)| = \delta^3(z_r - z'_{r'})\delta^3(z_i - z'_{i'}) \tag{3.151}$$

where

$$|\partial(q)/\partial(z)| = |\partial(q_r^{\,1},q_r^{\,2},q_r^{\,3},q_i^{\,1},\, q_i^{\,2},\, q_i^{\,3})/\partial(z_r^{\,1},z_r^{\,2},z_r^{\,3},z_i^{\,1},\, z_i^{\,2},\, z_i^{\,3})| = 1 \tag{3.152}$$

is the Jacobian of the transformation U. Thus the Fourier coefficients transform trivially under SU(3). For example,

$$b_C(q^*, s) \rightarrow b_C(z^*, s) \tag{3.153}$$

Since the integrand transforms as

$$\int d^3q_r d^3q_i \rightarrow \int d^3z_r d^3z_i\, |\partial(q)/\partial(z)| = \int d^3z_r d^3z_i \tag{3.154}$$

the wave function $\psi_C(t, \mathbf{x})$ transforms as an SU(3) scalar up to an inessential unitary transformation V of γ matrices: $\psi_C(t, \mathbf{x}) \rightarrow V\psi_C(t, \mathbf{x})$.[20]

3.5.1.5 Global SU(3) Spin ½ Complexon Fields

Having uncovered an SU(3) symmetry in the scalar field equations of Case 3A the generalization of the scalar field equations to the <u>3</u> representation of SU(3) is direct:

$$\psi_C^{\,a}(x) = \sum_{\pm s} \int d^3p_r d^3p_i\, N_C(p)\delta(\mathbf{p}_r\cdot\mathbf{p}_i/m^2)[b_C(p,a,s)u_C^{\,a}(p, s)e^{-i(p\cdot x + p^*\cdot x^*)/2} +$$
$$+ d_C^{\dagger}(p,a,s)v_C^{\,a}(p, s)e^{+i(p\cdot x + p^*\cdot x^*)/2}] \tag{3.155}$$

---

[20] The spinors $u_C(q^*,s)$ and $v_C(q^*,s)$ are unchanged up to a unitary transformation of the γ matrices ($V^{\dagger}\gamma'V = \gamma$). Thus the term $(U\mathbf{w})^*\cdot\gamma' = \mathbf{w}^*\cdot V\gamma V^{\dagger} \equiv \mathbf{w}^*\cdot\gamma$ in the expressions for the $u_C(q^*,s)$ and $v_C(q^*,s)$ spinors.

where $x = x_c$ for $a = 1, 2, 3$ with $u_C^a(p, s)$ and $v_C^a(p, s)$ being the product a spinor of type eq. 3.138 and a 3 element column vector c^a with b^{th} element

$$b^a(b) = \delta^{ab} \tag{3.156}$$

Under a global SU(3) transformation U the $\underline{3}$ complexon wave functions transform as

$$\psi_C'^a(x) = U^{ab}\psi_C^b(x) \tag{3.157}$$

In a subsequent discussion we will extend the global SU(3) symmetry described in these subsections to be color local SU(3) upon the introduction of the Yang-Mills color gluon interaction.

3.5.1.6 Lagrangian Formulation and Second Quantization of Complexons

In this subsection we will outline the canonical quantization of SU(3) singlet complexons with the quantum field equation

$$[i\gamma^0 \partial/\partial t + i\gamma \cdot (\nabla_r + i\nabla_i) - m]\psi_C(t, \mathbf{x_r}, \mathbf{x_i}) = 0 \tag{3.158}$$

and subsidiary condition

$$\nabla_r \cdot \nabla_i \, \psi_C(t, \mathbf{x_r}, \mathbf{x_i}) = 0 \tag{3.159}$$

We begin with the Lagrangian density

$$\mathcal{L} = \bar{\psi}_C(i\gamma^\mu D_\mu - m)\psi_C(x) \tag{3.160}$$

where $\bar{\psi}_C = \psi_C^\dagger \gamma^0$:

$$\psi_C^\dagger = [\psi_C(\mathbf{x_r}, \mathbf{x_i})]^\dagger \big|_{\mathbf{x_i} = -\mathbf{x_i}} \tag{3.161}$$

$$D_0 = \partial/\partial x^0$$
$$D_k = \partial/\partial x^k + i\,\partial/\partial x_i^k \tag{3.162}$$

with $x^k = x_r^k$ for $k = 1, 2, 3$. The invariant action (under real Lorentz transformations) is

$$I = \int d^7x \mathcal{L} \tag{3.163}$$

It is easy to show that the action is real

$$I^* = I \tag{3.164}$$

in a manner similar to the case considered in Appendix 2-A due to the form of $\psi_C^\dagger$ in eq. 3.161. (One has to change the integration over $\mathbf{x_i}$ to $-\mathbf{x_i}$ after taking the complex conjugate of I and performing manipulations similar to those in Appendix 3-A.)

The conjugate momentum is

$$\pi_{Ca} = \partial\mathcal{L}/\partial\dot{\psi}_{Ca} \equiv \partial\mathcal{L}/\partial(\partial\psi_{Ca}/\partial x^0) = i\psi_{C\,a}^{\dagger} \tag{3.165}$$

where a is a spinor index. It yields the non-zero anti-commutation relation

$$\{\psi_{C\,a}^{\dagger}(x),\ \psi_{Cb}(y)\} = \delta_{ab}\,\delta^3(x_r - y_r)\delta^3(x_i - y_i) \tag{3.166}$$

where x and y are complex. However we will see that the constraint eq. 3.159 is required. So the correct anti-commutator turns out to be

$$\{\psi_{C\,a}^{\dagger}(x),\ \psi_{Cb}(y)\} = -\delta_{ab}\delta'(\nabla_r\cdot\nabla_i/m^2)[\delta^3(x_r - y_r)\delta^3(x_i - y_i)] \tag{3.167}$$

where all ∇_r and ∇_i are ∇ (grad) derivatives with respect to x, and where $\delta'(\nabla_r\cdot\nabla_i)$ is the derivative of a delta function with the argument being differential operators such as those in eq. 3.159. The minus sign is due to the presence of a *derivative* of a delta-function and is not an issue.

The Hamiltonian density is

$$\mathcal{H} = \pi_C\dot{\psi}_C - \mathcal{L} = \psi_C^{\dagger}(-i\boldsymbol{\alpha}\cdot\mathbf{D} + \beta m)\psi_C \tag{3.168}$$

and the (unsymmetrized) energy-momentum tensor is

$$\mathcal{T}_{\mu\nu} = -g_{\mu\nu}\mathcal{L} + \partial\mathcal{L}/\partial(D^{\mu}\psi_C)D_{\nu}\psi_C \tag{3.169}$$

The conserved energy and momentum are

$$P^0 = H = \int d^3x_r d^3x_i\ \mathcal{T}^{00} = \int d^3x_r d^3x_i\ \mathcal{H} \tag{3.170}$$

and

$$P^i = \int d^3x_r d^3x_i\ \mathcal{T}^{0i} \tag{3.171}$$

We now proceed to establish the canonical anti-commutation relations. First, the second quantization of the complexon field uses the above Fourier coefficient anti-commutation relations (suitably rewritten):

$$\begin{aligned}
\{b_C(p,s),\ b_C^{\dagger}(p'^*,s')\} &= \delta_{ss'}\delta^3(\mathbf{p}_r - \mathbf{p'}_r)\delta^3(\mathbf{p}_i + \mathbf{p'}_i)\\
\{d_C(p,s),\ d_C^{\dagger}(p'^*,s')\} &= \delta_{ss'}\,\delta^3(\mathbf{p}_r - \mathbf{p'}_r)\delta^3(\mathbf{p}_i + \mathbf{p'}_i)\\
\{b_C(p,s),\ b_C(p'^*,s')\} &= \{d_C(p,s),\ d_C(p'^*,s')\} = 0\\
\{b_C^{\dagger}(p,s),\ b_C^{\dagger}(p'^*,s')\} &= \{d_C^{\dagger}(p,s),\ d_C^{\dagger}(p'^*,s')\} = 0\\
\{b_C(p,s),\ d_C^{\dagger}(p'^*,s')\} &= \{d_C(p,s),\ b_C^{\dagger}(p'^*,s')\} = 0
\end{aligned} \tag{3.172}$$

$$\{b_C^\dagger(p,s),\, d_C^\dagger(p'^*,s')\} = \{d_C(p,s),\, b_C(p'^*,s')\} = 0$$

The delta-function arguments $\delta^3(\mathbf{p_i} + \mathbf{p'_i})$ above have a positive sign in order to obtain $\delta^3(\mathbf{x_i} - \mathbf{y_i})$ in the field anti-commutator eq. 3.167.

The spinors, eq. 3.138, satisfy

$$\underset{\pm s}{\Sigma}\, u_\alpha(p,\,s)\bar{u}_\beta(p^*,\,s) = (2m)^{-1}(\not{p} + m)_{\alpha\beta} \tag{3.173}$$

$$\underset{\pm s}{\Sigma}\, v_\alpha(p,\,s)\bar{v}_\beta(p^*,\,s) = (2m)^{-1}(\not{p} - m)_{\alpha\beta}$$

remembering

$$\bar{u}_C(p^*,s) = w^{1T}(0)S_C(p)\gamma^0 = w^{1T}(0)[\cosh(\omega/2)I + \sinh(\omega/2)\gamma^0\boldsymbol{\gamma}\!\cdot\!\hat{\mathbf{w}}]\gamma^0 \tag{3.174}$$

by eqs. 3.137 since $\hat{\mathbf{w}}^{**} = \hat{\mathbf{w}}$.

We will now evaluate the equal-time anti-commutation relation using eqs. 3.136 and 3.137:

$$\{\psi_{C\,a}^\dagger(x),\, \psi_{Cb}(y)\} = \underset{\pm s,\, s'}{\Sigma}\int d^3p_r d^3p_i\, d^3p'_r d^3p'_i\, \delta(\mathbf{p_r}\!\cdot\!\mathbf{p_i}/m^2)\delta(\mathbf{p'_r}\!\cdot\!\mathbf{p'_i}/m^2)\, N_C(p')N_C(p)\cdot$$

$$\cdot[\{b_C^\dagger(p^*,s)u_{Ca}^\dagger(p^*,s)e^{+i(p\cdot x^* + p^*\cdot x)/2},\, b_C(p',s')u_{Cb}(p',s')e^{-i(p'\cdot y + p'^*\cdot\, y^*)/2}\}+$$

$$+ \{d_C(p^*,s)v_{Ca}^\dagger(p^*,s)e^{-i(p\cdot x^* + p^*\cdot x)/2},\, d_C^\dagger(p',s')v_{Cb}(p',s')e^{+i(p'\cdot y + p'^*\cdot\, y^*)/2}\}]$$

$$= \int d^3p_r d^3p_i\, N_C^2(p)[\delta(\mathbf{p_r}\!\cdot\!\mathbf{p_i}/m^2)]^2[((\not{p} + m)\gamma^0)_{ba}\,e^{+i(p\cdot x^* + p^*\cdot x)/2 - i(p^*\cdot y + p\cdot y^*)/2} +$$

$$+((\not{p} - m)\gamma^0)_{ba}\,e^{-i(p\cdot x^* + p^*\cdot x)/2 + i(p^*\cdot y + p\cdot y^*)/2}]/(2m)$$

Next we use eq. 3.140 and the identity

$$[\delta(x - y)]^2 = -\tfrac{1}{2}\,\delta'(x - y) \equiv -\tfrac{1}{2}\, d\delta(x - y)/dx \tag{3.175}$$

which can be derived from the step function identity $\theta(x - y) = [\theta(x - y)]^2$ to obtain

$$\{\psi_{C\,a}^\dagger(x),\psi_{Cb}(y)\} = -\tfrac{1}{2}\int d^3p_r d^3p_i N_C^2(p)\delta'(\mathbf{p_r}\!\cdot\!\mathbf{p_i}/m^2)[((\not{p}+m)\gamma^0)_{ba}\,e^{-ipr\cdot(xr - yr) + ipi\cdot(xi - yi)} +$$

$$+ ((\not{p} - m)\gamma^0)_{ba}\,e^{+ipr\cdot(xr - yr) - ipi\cdot(xi - yi)}]/(2m)$$

$$= -\tfrac{1}{2}\delta_{ba}\int d^3p_r d^3p_i N_C^2(p)\delta'(\mathbf{p_r}\!\cdot\!\mathbf{p_i}/m^2)p^0 e^{-ipr\cdot(xr - yr) + ipi\cdot(xi - yi)}/m$$

$$= -\,\delta_{ab}\, \delta'(\nabla_r\!\cdot\!\nabla_i/m^2)[\delta^3(\mathbf{x_r} - \mathbf{y_r})\delta^3(\mathbf{x_i} - \mathbf{y_i})] \tag{3.176}$$

The grad operators, ∇_r and ∇_i, are derivatives are with respect to x in the Dirac delta functions. The factor[21] $\delta'(\nabla_r \cdot \nabla_i)$ expresses the orthogonality constraint in coordinate space on the momenta. It is analogous to the transversality constraint on the electromagnetic vector potential commutator:

$$[\pi_A^{\,j}(x), A_k(y)] = -i\,\delta^{tr}_{\;jk}(x-y) \tag{3.177}$$

$$\delta^{tr}_{\;jk}(x-y) = (\delta_{jk} - \partial_j\partial_k/\nabla^2)\,\delta^3(x-y) \tag{3.178}$$

where $\partial_k = \partial/\partial x_k$.

3.5.1.7 Complexon Feynman Propagator

The complexon Feynman propagator for ψ_C is[22]

$$iS_C(x, y) = \theta(x^0 - y^0)<0|\psi_C(x)\psi_C^\dagger(y)\gamma^0|0> - \theta(y^0 - x^0)<0|\psi_C^\dagger(y)\gamma^0\psi_C(x)|0> \tag{3.179}$$

$$= \int d^3p_r d^3p_i N_C^2(p)[\delta(\mathbf{p_r}\cdot\mathbf{p_i}/m^2)]^2 \{\theta(x^0-y^0)(\not{p}+m)e^{-i(p^*\cdot(x-y)+p\cdot(x^*-y^*))/2} -$$
$$- \theta(y^0-x^0)(\not{p}-m)e^{+i(p^*\cdot(x-y)+p\cdot(x^*-y^*))/2}\}/(2m)$$

$$= -(4\pi)^{-1}\int dp^0 d^3p_r d^3p_i(2\pi)^{-6}\delta'(\mathbf{p_r}\cdot\mathbf{p_i}/m^2)(\not{p}+m)e^{-i(p^*\cdot(x-y)+p\cdot(x^*-y^*))/2}/(p^2-m^2+i\epsilon)$$

$$= -\tfrac{1}{2}\int dp^0 d^3p_r d^3p_i\,\delta'(\mathbf{p_r}\cdot\mathbf{p_i}/m^2)(\not{p}+m)(2\pi)^{-7}\exp[-ip^0(x^0-y^0)+$$
$$+ i\mathbf{p_r}\cdot(\mathbf{x_r}-\mathbf{y_r}) - i\mathbf{p_i}\cdot(\mathbf{x_i}-\mathbf{y_i})]/(p^2-m^2+i\epsilon) \tag{3.180}$$

The integral can be written in the form:

$$I = \int dp^0 d^3p_r d^3p_i\delta'(\mathbf{p_r}\cdot\mathbf{p_i}/m^2)(\not{p}+m)\exp[-ip^0(x^0-y^0)+i\mathbf{p_r}\cdot(\mathbf{x_r}-\mathbf{y_r})-i\mathbf{p_i}\cdot(\mathbf{x_i}-\mathbf{y_i})]/(p^2-m^2+i\epsilon)$$
$$= \int d^4p_r dM^2\delta'(\nabla_r\cdot\nabla_i/m^2)(p^0\gamma^0-(\mathbf{p_r}-\nabla_i)\cdot\gamma+m)\exp[-ip^0(x^0-y^0)+i\mathbf{p_r}\cdot(\mathbf{x_r}-\mathbf{y_r})]\cdot$$
$$\cdot J(\mathbf{x_i}-\mathbf{y_i},M^2)/(p_r^2-M^2+i\epsilon) \tag{3.181}$$

where $p_r^2 = p^{0\,2} - \mathbf{p_r}\cdot\mathbf{p_r}$ and

$$J(\mathbf{x_i}-\mathbf{y_i}, M^2) = (2\pi)^{-3}\int d^3p_i\delta(M^2+\mathbf{p_i}^2-m^2)\exp[-i\mathbf{p_i}\cdot(\mathbf{x_i}-\mathbf{y_i})] \tag{3.182}$$
$$= (2\pi)^{-2}|\mathbf{x_i}-\mathbf{y_i}|^{-1}\theta(m^2-M^2)\sin((m^2-M^2)^{1/2}|\mathbf{x_i}-\mathbf{y_i}|)$$

The complexon Feynman propagator can be rearranged into the form of a spectral integral:

---

[21] A derivative of a delta function containing grad operators.

[22] The reader, upon seeing the additional integrations $\int d^3p_i$ might suspect that they would ultimately lead to divergence issues in perturbation theory calculations. However the $\delta'(\mathbf{p_r}\cdot\mathbf{p_i}/m^2)$ term compensates in part for the additional integrations by four powers of momentum since $\delta'(\mathbf{p_r}\cdot\mathbf{p_i}/m^2) = (|\mathbf{p_r}||\mathbf{p_i}|/m^2)^{-2}\delta'(\cos\theta_{ri})$ where θ_{ri} is the angle between the momenta. As a result only 2 fermion and 3 fermion loop integrations would potentially have difficulties if one uses the conventional approach to perturbation theory. If one uses the approach of Blaha (2003) and (2005a) then there are no divergences.

$$iS_C(x, y) = -\int dM\, (i\gamma^0 \partial/\partial x^0 - i(\nabla_r - i\nabla_i)\cdot\gamma + m)\delta'(\nabla_r\cdot\nabla_i/m^2)J(\mathbf{x_i} - \mathbf{y_i}, M^2)\triangle_F(x - y, M)$$

$$(3.183)$$

where

$$\triangle_F(x - y, M) = (2\pi)^{-4}\int d^4 p_r \exp[-ip^0(x^0 - y^0) + i\mathbf{p_r}\cdot(\mathbf{x_r} - \mathbf{y_r})]/(p_r^2 - M^2 + i\varepsilon)$$

$$(3.184)$$

3.5.1.8 Case 4: Left-handed Tachyon Complexons

In this case $\mathbf{\hat{u}_r}\cdot\mathbf{\hat{u}_i} = 0$ again. However we add an imaginary term to ω to obtain a manifest Left-handed L_C boost[23]

$$\Lambda_{CL}(\mathbf{v_c}) = \exp[i(\omega + i\pi/2)\mathbf{\hat{w}}\cdot\mathbf{K}]$$

$$(3.185)$$

where ω remains

$$\omega = (\omega_r^2 - \omega_i^2)^{\frac{1}{2}}$$

$$(3.186)$$

and

$$\mathbf{\hat{w}} = (\omega_r\mathbf{\hat{u}_r} + i\omega_i\mathbf{\hat{u}_i})/\omega$$

$$(3.187)$$

$$\mathbf{\hat{w}}\cdot\mathbf{\hat{w}} = \mathbf{\hat{u}_r}\cdot\mathbf{\hat{u}_r} = \mathbf{\hat{u}_i}\cdot\mathbf{\hat{u}_i} = 1$$

$$(3.188)$$

$$\mathbf{v_c} = \mathbf{\hat{w}}\tanh(\omega + i\pi/2) = \mathbf{\hat{w}}\cotanh(\omega)$$

$$(3.189)$$

Letting $\omega_L = \omega + i\pi/2$ we find, as before,

$$\cosh(\omega_L) = i\sinh(\omega) = -\gamma = i\gamma_s$$

$$(3.190)$$

$$\sinh(\omega_L) = i\cosh(\omega) = -\beta\gamma = i\beta\gamma_s$$

with, $\beta = v_c = |\mathbf{v_c}| > 1$, $\gamma_s = (\beta^2 - 1)^{-\frac{1}{2}}$, and

$$\sinh(\omega) = \gamma_s$$

$$(3.191)$$

$$\cosh(\omega) = \beta\gamma_s$$

Thus we denote $\Lambda_{CL}(\mathbf{v_c})$ by

$$\Lambda_{CL}(\mathbf{v_c}) \equiv \Lambda_{CL}(\omega, \mathbf{\hat{w}})$$

$$(3.192)$$

The corresponding spinor boost transformation is:

$$S_{CL}(\Lambda_{CL}(\omega, \mathbf{\hat{w}})) = \exp(-i\omega_L\sigma_{0i}\hat{w}_i/2) = \exp(-\omega_L\gamma^0\gamma\cdot\mathbf{\hat{w}}/2)$$

$$= \cosh(\omega_L/2)I + \sinh(\omega_L/2)\gamma^0\gamma\cdot\mathbf{\hat{w}}$$

$$(3.193)$$

The momentum space equation generated by $S_{CL}(\Lambda_{CL}(\omega, \mathbf{\hat{w}}))$ is

$$\{m\cosh(\omega_L)\gamma^0 - m\sinh(\omega_L)\gamma\cdot(\omega_r\mathbf{\hat{u}_r} + i\omega_i\mathbf{\hat{u}_i})/\omega - m\}e^{+ip\cdot x}w_{cL}(p) = 0 \quad (3.194)$$

or

$$\{im\sinh(\omega)\gamma^0 - im\cosh(\omega)\gamma\cdot(\omega_r\mathbf{\hat{u}_r} + i\omega_i\mathbf{\hat{u}_i})/\omega - m\}e^{+ip\cdot x}w_{cL}(p) = 0 \quad (3.195)$$

---

[23] The reader can readily verify the form is consistent that generated by an L_C boost transformation.

where $p \cdot x = Et - \mathbf{p} \cdot \mathbf{x}$ after performing a corresponding left-handed superluminal coordinate transformation in the exponential factor. Thus the positive energy wave function is transformed into a negative energy wave function by the transformation.

The momentum 4-vector is defined by

$$p = (p^0, \mathbf{p}) \tag{3.196}$$

where

$$p^0 = m \sinh(\omega) \qquad\qquad \mathbf{p} = \mathbf{p}_r + i\mathbf{p}_i \tag{3.197}$$

with

$$\mathbf{p}_r = m\omega_r \hat{\mathbf{u}}_r \cosh(\omega)/\omega \quad \mathbf{p}_i = m\omega_i \hat{\mathbf{u}}_i \cosh(\omega)/\omega \tag{3.198}$$

and

$$\mathbf{p}_r \cdot \mathbf{p}_i = 0 \tag{3.199}$$

then eq. 3.195 becomes the complexon tachyon equation

$$[i p \cdot \gamma - m] e^{+i p \cdot x} w_{cL}(p) = 0 \tag{3.200}$$

with a complex 3-momentum $\mathbf{p}$ and the tachyon 4-momentum mass shell condition:[24]

$$p^2 = p^{0\,2} - \mathbf{p}_r^2 + \mathbf{p}_i^2 = -m^2 \tag{3.201}$$

Eq. 3.200 is the momentum space equivalent of the wave equation

$$[\gamma^0 \partial/\partial t + \gamma \cdot (\nabla_r + i\nabla_i) - m]\psi_{CL}(t, \mathbf{x}_r, \mathbf{x}_i) = 0 \tag{3.202}$$

or

$$[\gamma \cdot \nabla - m]\psi_{CL}(t, \mathbf{x}_r, \mathbf{x}_i) = 0 \tag{3.203}$$

with the subsidiary condition on the wave function

$$\nabla_r \cdot \nabla_i \, \psi_{CL}(t, \mathbf{x}_r, \mathbf{x}_i) = 0 \tag{3.204}$$

also holds. We note that eq. 3.202 is covariant under the real Lorentz group and eq. 3.204 can be easily put into (real Lorentz group) covariant form.

Before considering a Lagrangian formulation and the Fourier operator representation of $\psi_{CL}(t, \mathbf{x}_r, \mathbf{x}_i)$ we will define the tachyon spinors, and its associated real and imaginary spin operators.

The spinor generated from a spin up Dirac spinor at rest by the L_C spinor boost eq. 3.193 is

$$w_{cL}(p) = S_{CL} w(0) = [\cosh(\omega_L/2)I + \sinh(\omega_L/2)\gamma^0 \gamma \cdot \hat{\mathbf{w}}]w(0) \tag{3.205}$$

---

[24] Note that the presence of the $\mathbf{p}_i^2$ term does not change the tachyon requirement that $\mathbf{p}_r^2 \geq m^2$ as seen in the previous cases.

Following a procedure similar to Appendix 3-A (which the reader may wish to examine first) we define four spinors for Dirac particles at rest with eq. 3-A.2. Then by applying a boost to these rest spinors we find the L_C tachyon spinors:

$$S_{CL}w^k(0) = w_{cL}{}^k(p) \qquad (3.206)$$

and from these tachyon spinors we generalize to tachyon spinors $u_{CL}(p, s)$ and $v_{CL}(p, s)$ in a manner similar to that of the previous case.

Eqs. 3.200 through 3.204 imply that the wave function solution of eq. 3.200, subject to the subsidiary condition eq. 3.204, has the form

$$\psi_{CL}(x) = \sum_{\substack{\pm s \\ p_r^2 \geq m^2}} \int d^3p_r d^3p_i \, N_{CL}(p)\delta(\mathbf{p_r \cdot p_i}/m^2)[b_{CL}(p,s)u_{CL}(p, s)e^{-i(p\cdot x + p^*\cdot x^*)/2} +$$
$$+ d_{CL}{}^\dagger(p,s)v_{CL}(p, s)e^{+i(p\cdot x + p^*\cdot x^*)/2}] \qquad (3.207)$$

where $\mathbf{p} = \mathbf{p_r} + i\mathbf{p_i}$, $\mathbf{x} = \mathbf{x_r} - i\mathbf{x_i}$, $p\cdot x = p^0x^0 - \mathbf{p\cdot x}$, and $b_{CL}(p, s)$ and $d_{CL}(p,s)$ are tachyon Fourier coefficients.

3.5.1.9 Global SU(3) Symmetry

We can show that there is also a global SU(3) symmetry present here as shown in the previous case. The demonstration is similar to that of eqs. 3.143 – 3.156.

3.5.1.10 Light-Front Quantization of Tachyonic Complexons

Because of the momentum constraint $\mathbf{p_r}^2 \geq m^2$ the set of solutions of the form of eq. 3.207 is incomplete and the result of second quantization would not be an equal time anti-commutator expression consisting of derivatives of delta functions (eq. 3.176) but rather an analogue to previous unsuccessful attempts to create a second quantized tachyon theory.[25]

Therefore we will use light-front coordinates, and left and right handed field operators (as previously) to obtain a successful second quantization of this new type of tachyon.

The "missing" factor of i in the first term of eq. 3.203 requires the Lagrangian to be different from the conventional Dirac Lagrangian in order for the Lagrangian to be real. The simplest, physically acceptable, free spin ½ tachyon Lagrangian density for ψ_{CL} is:

$$\mathcal{L}_{CL} = \psi_{CL}{}^C(x)(\gamma\cdot\nabla - m)\psi_{CL}(x) \qquad (3.208)$$

where

$$\psi_{CL}{}^C(x) = [\psi_{CL}(x)]^\dagger|_{\mathbf{x_i} = -\mathbf{x_i}} \, i\gamma^0\gamma^5 \qquad (3.209)$$

---

[25] Such as G. Feinberg, Phys. Rev. **159**, 1089 (1967).

is similar to eq. 3.161. In words, eq. 3.209 states: take the Hermitean conjugate of $\psi_{CL}(x)$; change x_i to $-x_i$; and then post-multiply by the indicated factors.

The free complexon invariant action (under real Lorentz transformations) is

$$I = \int d^7x \, \mathcal{L}_{CL} \tag{3.210}$$

The action can be shown to be real

$$I^* = I \tag{3.211}$$

in a manner similar to the case considered in Appendix 3-A. The tachyonic complexon's energy-momentum tensor is

$$\mathcal{T}_{CL\mu\nu} = - g_{\mu\nu} \mathcal{L}_{CL} + \partial \mathcal{L}_{CL}/\partial(D^\mu \psi_{CL}) \, D_\nu \psi_{CL}$$
$$= i \psi_{CL}{}^C \gamma^0 \gamma^5 \gamma_\mu D_\nu \psi_{CL} \tag{3.212}$$

where

$$D_0 = \partial/\partial x^0$$
$$D_k = \partial/\partial x_r{}^k + i \, \partial/\partial x_i{}^k \tag{3.213}$$

and thus the conserved energy and momentum are

$$P^0 = H = \int d^3x_r d^3x_i \, \mathcal{T}_{CL}{}^{00} = i\int d^3x_r d^3x_i \psi_{CL}{}^C \gamma^5 (\boldsymbol{\alpha} \cdot \mathbf{D} + \beta m) \psi_{CL} \tag{3.214}$$
$$P^k = \int d^3x_r d^3x_i \, \mathcal{T}_{CL}{}^{0k} = - i\int d^3x_r d^3x_i \, \psi_{CL}{}^C \gamma^5 D^k \psi_{CL} \tag{3.215}$$

Having defined a suitable tachyon Lagrangian we can now proceed to its canonical quantization. The conjugate momentum can be calculated from the Lagrangian density eq. 3.212:

$$\pi_{CLa} = \partial \mathcal{L}_{CL}/\partial \dot{\psi}_{CLa} \equiv \partial \mathcal{L}_{CL}/\partial(\partial \psi_{CLa}/\partial t) = -i([\psi_{CL}(x)]^\dagger|_{x_i = -x_i} \gamma^5)_a \tag{3.216}$$

The resulting non-zero, canonical anti-commutation relations are

$$\{\pi_{CLa}(x), \psi_{CLb}(y)\} = i \, \delta_{ab} \, \delta^3(x_r - y_r) \delta^3(x_i - y_i)$$

based on locality in both real and imaginary coordinates:

$$\{\psi_{CL}{}^\dagger{}_a(x)|_{x_i = -x_i}, \psi_{Tb}(y)\} = - [\gamma^5]_{ab} \, \delta^3(x_r - y_r) \delta^3(x_i - y_i) \tag{3.217}$$

At this point we might attempt to complete the canonical quantization procedure in the conventional manner by Fourier expanding the field and specifying anti-commutation relations for the Fourier component amplitudes. However the incompleteness of the set of plane waves, which are limited by the restriction $p_r^2 \geq m^2$, causes the equal time anti-commutator of the fields *not* to yield a δ-functions.

Therefore we turn to the previous successful approach to tachyon quantization[26] and decompose the tachyonic complexon field into left-handed and right-handed parts and then second quantize in light-front coordinates.

3.5.2 Separation into Left-Handed and Right-Handed Fields

As before we will use a transformed set of Dirac matrices to develop our left-handed and right-handed tachyon formulations. The γ^5 chirality operator's eigenvalues define handedness: +1 corresponds to right-handed; and -1 corresponds to left-handed:

$$\gamma^5 \psi_{CLL} = -\psi_{CLL} \qquad\qquad \gamma^5 \psi_{CLR} = \psi_{CLR} \qquad\qquad (3.218)$$

We define left-handed and right-handed tachyon fields with the projection operators:

$$\begin{aligned}
C^{\pm} &= \tfrac{1}{2}(I \pm \gamma^5) \\
C^{+} + C^{-} &= I \\
C^{\pm\,2} &= C^{\pm} \\
C^{+}C^{-} &= 0
\end{aligned} \qquad\qquad (3.219)$$

with the result

$$\begin{aligned}
\psi_{CLL} &= C^{-}\psi_{CL} \\
\psi_{CLR} &= C^{+}\psi_{CL}
\end{aligned} \qquad\qquad (3.220)$$

We can calculate the commutation relations of the left-handed and right-handed tachyonic complexon fields from eq. 3.217 by pre-multiplying and post-multiplying by $\tfrac{1}{2}(1 - \gamma^5)$ and $\tfrac{1}{2}(1 + \gamma^5)$. The results are:

$$\{\psi_{CLLa}^{\dagger}(x)\big|_{\mathbf{x_i} = -\mathbf{x_i}},\ \psi_{CLLb}(y)\} = C^{-}_{ab}\,\delta^6(x - y) \qquad\qquad (3.221)$$

$$\{\psi_{CLRa}^{\dagger}(x)\big|_{\mathbf{x_i} = -\mathbf{x_i}},\ \psi_{CLRb}(y)\} = -C^{+}_{ab}\,\delta^6(x - y) \qquad\qquad (3.222)$$

$$\{\psi_{CLLa}^{\dagger}(x)\big|_{\mathbf{x_i} = -\mathbf{x_i}},\ \psi_{CLRb}(y)\} = \{\psi_{CLRa}^{\dagger}(x)\big|_{\mathbf{x_i} = -\mathbf{x_i}},\ \psi_{CLLb}(x')\} = 0 \qquad\qquad (3.223)$$

where

$$\delta^6(x - y) = \delta^3(x_r - y_r)\delta^3(x_i - y_i) \qquad\qquad (3.224)$$

The Lagrangian density of eq. 3.208 decomposes into left-handed and right-handed parts: (The change x_i to $-x_i$ will be understood in $\psi_{CLL}^{\dagger}(x)$ and $\psi_{CLR}^{\dagger}(x)$ in the following.)

---

[26] Blaha (2006) discusses this case in detail.

$$\mathcal{L}_{CL} = \psi_{CLL}{}^{\dagger}\gamma^0 i\gamma^\mu \partial_\mu \psi_{CLL} - \psi_{CLR}{}^{\dagger}\gamma^0 i\gamma^\mu \partial_\mu \psi_{CLR} - im[\psi_{CLR}{}^{\dagger}\gamma^0 \psi_{CLL} - \psi_{CLL}{}^{\dagger}\gamma^0 \psi_{CLR}]$$

$$(3.225)$$

3.5.3 Further Separation into + and – Light-Front Complexon Fields

As previously, we now use light-front coordinates and quantization to obtain a successful second quantization of this form of tachyon field. Light-front variables, in the present case where we have to contend with complex 3-vectors, are defined by real coordinates and derivatives:

$$x^\pm = (x^0 \pm x_r{}^3)/\sqrt{2}$$

$$\partial/\partial x^\pm \equiv \partial^\mp \equiv (\partial/\partial x^0 \pm \partial/\partial x_r{}^3)/\sqrt{2}$$

$$(3.226)$$

with the "transverse" real coordinate variables, $x_r{}^1$ and $x_r{}^2$, and imaginary coordinate variables $x_i{}^1$, $x_i{}^2$, and $x_i{}^3$.

The inner product of two 4-vectors has the form

$$x{\cdot}y = x^+y^- + y^+x^- + i[y_i{}^3(x^+ - x^-) + x_i{}^3(y^+ - y^-)]/\sqrt{2} + x_i{}^3 y_i{}^3 - (\mathbf{x}_{r\perp} - i\mathbf{x}_{i\perp})\cdot(\mathbf{y}_{r\perp} - i\mathbf{y}_{i\perp})$$

$$(3.227)$$

with

$$\mathbf{x}_{r\perp} = (x_r{}^1, x_r{}^2) \qquad \mathbf{x}_{i\perp} = (x_i{}^1, x_i{}^2)$$

$$\mathbf{y}_{r\perp} = (y_r{}^1, y_r{}^2) \qquad \mathbf{y}_{i\perp} = (y_i{}^1, y_i{}^2)$$

$$(3.228)$$

where $x = (x^0, \mathbf{x} = \mathbf{x}_r - i\mathbf{x}_i)$ and $y = (y^0, \mathbf{y} = \mathbf{y}_r - i\mathbf{y}_i)$. Momenta are always defined as $p = (p^0, \mathbf{p} = \mathbf{p}_r + i\mathbf{p}_i)$.

The light-front definition of Dirac matrices is:

$$\gamma^\pm = (\gamma^0 \pm \gamma^3)/\sqrt{2}$$

$$(3.229)$$

with transverse matrices γ^1 and γ^2 defined as usual. Note:

$$\gamma^{\pm 2} = 0$$

We define "+" and "–" tachyon fields with the projection operators:

$$R^\pm = \tfrac{1}{2}(I \pm \gamma^0\gamma^3)$$

$$(3.230)$$

Left-handed, ± light-front fields: $\qquad \psi_{CLL}{}^\pm = R^\pm C^- \psi_{CL}$ $\qquad (3.231)$

Right-handed, ± light-front fields: $\qquad \psi_{CLR}{}^\pm = R^\pm C^+ \psi_{CL}$

Transforming to light-front variables and fields as above we obtain the light-front free tachyon Lagrangian:

$$\mathcal{L}_{CL} = 2^{\frac{1}{2}}\psi_{CLL}^{++\dagger}i\partial^{-}\psi_{CLL}^{+} + 2^{\frac{1}{2}}\psi_{CLL}^{-\dagger}i\partial^{+}\psi_{CLL}^{-} - \psi_{CLL}^{++\dagger}\gamma^{0}[i\gamma_{\perp}\cdot\nabla_{r\perp} - \gamma\cdot\nabla_{i}]\psi_{CLL}^{-} -$$
$$- \psi_{CLL}^{-\dagger}\gamma^{0}[i\gamma_{\perp}\cdot\nabla_{r\perp} - \gamma\cdot\nabla_{i}]\psi_{CLL}^{+} - 2^{\frac{1}{2}}\psi_{CLR}^{++\dagger}i\partial^{-}\psi_{CLR}^{+} - 2^{\frac{1}{2}}\psi_{CLR}^{-\dagger}i\partial^{+}\psi_{CLR}^{-} +$$
$$+ \psi_{CLR}^{++\dagger}\gamma^{0}[i\gamma_{\perp}\cdot\nabla_{r\perp} - \gamma\cdot\nabla_{i}]\psi_{CLR}^{-} + \psi_{CLR}^{-\dagger}\gamma^{0}[i\gamma_{\perp}\cdot\nabla_{r\perp} - \gamma\cdot\nabla_{i}]\psi_{CLR}^{+} -$$
$$- im[\psi_{CLR}^{++\dagger}\gamma^{0}\psi_{CLL}^{-} - \psi_{CLL}^{++\dagger}\gamma^{0}\psi_{CLR}^{-} + \psi_{CLR}^{-\dagger}\gamma^{0}\psi_{CLL}^{+} - \psi_{CLL}^{-\dagger}\gamma^{0}\psi_{CLR}^{+}]$$

$$(3.232)$$

(Note the similarity to the previous tachyon case.) Again the difference in signs between the left-handed and right-handed terms will be a crucial factor in the derivation of the left-handed features of the Standard Model.

Eq. 3.232 generates the equations of motion:

$$2^{\frac{1}{2}}i\partial^{-}\psi_{CLL}^{+} - \gamma^{0}[i\gamma_{\perp}\cdot\nabla_{r\perp} - \gamma\cdot\nabla_{i}]\psi_{CLL}^{-} + im\gamma^{0}\psi_{CLR}^{-} = 0 \qquad (3.233)$$
$$2^{\frac{1}{2}}i\partial^{-}\psi_{CLR}^{+} - \gamma^{0}[i\gamma_{\perp}\cdot\nabla_{r\perp} - \gamma\cdot\nabla_{i}]\psi_{CLR}^{-} + im\gamma^{0}\psi_{CLL}^{-} = 0$$
$$2^{\frac{1}{2}}i\partial^{+}\psi_{CLL}^{-} - \gamma^{0}[i\gamma_{\perp}\cdot\nabla_{r\perp} - \gamma\cdot\nabla_{i}]\psi_{CLL}^{+} + im\gamma^{0}\psi_{CLR}^{+} = 0$$
$$2^{\frac{1}{2}}i\partial^{+}\psi_{CLR}^{-} - \gamma^{0}[i\gamma_{\perp}\cdot\nabla_{r\perp} - \gamma\cdot\nabla_{i}]\psi_{CLR}^{+} + im\gamma^{0}\psi_{CLL}^{+} = 0$$

Eqs. 3.233 show that ψ_{CLL}^{-} and ψ_{CLR}^{-} are dependent fields that are functions of ψ_{CLL}^{+} and ψ_{CLR}^{+} on the light-front where x^{+} equals a constant. They can be expressed in an integral form as well. (The independent fields ψ_{CLL}^{+} and ψ_{CLR}^{+} play a fundamental role in tachyonic complexon theory and are used to define "in" and "out" tachyon states in perturbation theory.)

The conjugate momenta implied by eq. 3.232 are

$$\pi_{CLL}^{+} = \partial\mathcal{L}/\partial(\partial^{-}\psi_{CLL}^{+}) = 2^{\frac{1}{2}}i\psi_{CLL}^{++\dagger} \qquad (3.234)$$
$$\pi_{CLL}^{-} = \partial\mathcal{L}/\partial(\partial^{-}\psi_{CLL}^{-}) = 0$$
$$\pi_{CLR}^{+} = \partial\mathcal{L}/\partial(\partial^{-}\psi_{CLR}^{+}) = -2^{\frac{1}{2}}i\psi_{CLR}^{++\dagger} \qquad (3.235)$$
$$\pi_{CLR}^{-} = \partial\mathcal{L}/\partial(\partial^{-}\psi_{CLR}^{-}) = 0$$

x^{+} plays the role of the "time" variable in light-front quantized theories. So we define canonical equal x^{+} anti-commutation relations for spin ½ tachyonic complexons also.

The canonical equal-light-front $(x^{+} = y^{+})$ anti-commutation relations of the independent fields would normally be:

$$\{\psi_{CLL}^{++\dagger}{}_{a}(x),\psi_{CLL}^{+}{}_{b}(y)\} = 2^{-1}[C^{-}R^{+}]_{ab}\delta(x^{-} - y^{-})\delta^{2}(x_{r} - y_{r})\delta^{3}(x_{I} - y_{i}) \quad (3.236)$$
$$\{\psi_{CLR}^{++\dagger}{}_{a}(x), \psi_{CLR}^{+}{}_{b}(y)\} = -2^{-1}[C^{+}R^{+}]_{ab}\delta(x^{-} - y^{-})\delta^{2}(x_{r} - y_{r})\delta^{3}(x_{I} - y_{i}) \quad (3.237)$$
$$\{\psi_{CLL}^{+}{}_{a}{}^{\dagger}(x), \psi_{CLR}^{+}{}_{b}(y)\} = \{\psi_{CLR}^{+}{}_{a}{}^{\dagger}(x), \psi_{CLL}^{+}{}_{b}(y)\} = 0 \quad (3.238)$$
$$\{\psi_{CLL}^{+}{}_{a}(x), \psi_{CLR}^{+}{}_{b}(y)\} = \{\psi_{CLR}^{+}{}_{a}{}^{\dagger}(x), \psi_{CLL}^{++}{}_{b}(y)\} = 0 \quad (3.239)$$

As in the previous case they will be modified.

Again we see that the right-handed tachyon anti-commutation relation (eq. 3.237) has a minus sign relative to the corresponding conventional right-handed anti-commutation relation.

The sign differences between the left-handed and right-handed Lagrangian terms ultimately lead to parity violating features in the Standard Model Lagrangian.

3.5.3.1 Left-Handed Tachyonic Complexons

The free, "+" light-front, left-handed tachyonic complexon Fourier expansion is:

$$\psi_{CLL}^{+}(x_r, x_i) = \sum_{\pm s} \int d^2p_r dp^+ d^3p_i \, N_{CLL}^{+}(p)\theta(p^+)\delta((p_i^3(p^+ - p^-)/\sqrt{2} + \mathbf{p}_{r\perp}\cdot\mathbf{p}_{i\perp})/m^2)\cdot$$

$$\cdot[b_{CLL}^{+}(p, s)u_{CLL}^{+}(p, s)e^{-i(p\cdot x + p^*\cdot x^*)/2} + d_{CLL}^{+\dagger}(p, s)v_{CLL}^{+}(p, s)e^{+i(p\cdot x + p^*\cdot x^*)/2}]$$

$$(3.240)$$

Its Hermitean conjugate is

$$\psi_{CLL}^{+\dagger}(x_r, x_i) = \sum_{\pm s} \int d^2p_r dp^+ d^3p_i \, N_{CLL}^{+}(p)\theta(p^+)\delta((p_i^3(p^+ - p^-)/\sqrt{2} + \mathbf{p}_{r\perp}\cdot\mathbf{p}_{i\perp})/m^2)\cdot$$

$$\cdot[b_{CLL}^{\dagger}(p^*,s)u_{CLL}^{\dagger}(p^*,s)e^{+i(p^*\cdot x + p\cdot x^*)/2} + d_{CLL}(p^*,s)v_{CLL}^{\dagger}(p^*,s)e^{-i(p^*\cdot x + p\cdot x^*)/2}]$$

$$(3.241)$$

where $\mathbf{p} = \mathbf{p}_r + i\mathbf{p}_i$, $\mathbf{x} = \mathbf{x}_r - i\mathbf{x}_i$, $p\cdot x = p^0x^0 - \mathbf{p}\cdot\mathbf{x}$, and † indicates Hermitean conjugate. The spinors are

$$u_{CLL}^{+}(p, s) = C^- R^+ S_{CL}w^1(0)$$
$$u_{CLL}^{+}(p, -s) = C^- R^+ S_{CL}w^2(0)$$
$$v_{CLL}^{+}(p, s) = C^- R^+ S_{CL}w^3(0)$$
$$v_{CLL}^{+}(p, -s) = C^- R^+ S_{CL}w^4(0)$$
$$u_{CLL}^{+\dagger}(p^*, s) = w^{1T}(0)S_{CL}R^+C^-$$
$$u_{CLL}^{+\dagger}(p^*, -s) = w^{2T}(0)S_{CL}R^+C^-$$
$$v_{CLL}^{+\dagger}(p^*, s) = w^{3T}(0)S_{CL}R^+C^-$$
$$v_{CLL}^{+\dagger}(p^*, -s) = w^{4T}(0)S_{CL}R^+C^-$$

$$(3.242)$$

where the superscript "T" indicates the transpose (These spinors are described in Appendix 3-A.) and

$$N_{CLL}^{+}(p) = (2\pi)^{-3}(2m/p^+)^{\frac{1}{2}}$$

$$(3.243)$$

The anti-commutation relations of the Fourier coefficient operators are

$$\{b_{CLL}(p,s), b_{CLL}^{\dagger}(p'^*,s')\} = 2^{-\frac{1}{2}}\delta_{ss'}\delta(p^+ - p'^+)\delta^2(\mathbf{p}_r - \mathbf{p}'_r)\delta^3(\mathbf{p}_i + \mathbf{p}'_i)$$
$$\{d_{CLL}(p,s), d_{CLL}^{\dagger}(p'^*,s')\} = 2^{-\frac{1}{2}}\delta_{ss'}\delta(p^+ - p'^+)\delta^2(\mathbf{p}_r - \mathbf{p}'_r)\delta^3(\mathbf{p}_i + \mathbf{p}'_i)$$
$$\{b_{CLL}(p,s), b_{CLL}(p'^*,s')\} = \{d_{CLL}(p,s), d_{CLL}(p'^*,s')\} = 0$$
$$\{b_{CLL}^{\dagger}(p,s), b_{CLL}^{\dagger}(p'^*,s')\} = \{d_{CLL}^{\dagger}(p,s), d_{CLL}^{\dagger}(p'^*,s')\} = 0$$

$$(3.244)$$

$$\{b_{CLL}(p,s), d_{CLL}^{\dagger}(p'^*,s')\} = \{d_{CLL}(p,s), b_{CLL}^{\dagger}(p'^*,s')\} = 0$$
$$\{b_{CLL}^{\dagger}(p,s), d_{CLL}^{\dagger}(p'^*,s')\} = \{d_{CLL}(p,s), b_{CLL}(p'^*,s')\} = 0$$

The delta-function arguments $\delta^3(\mathbf{p_i} + \mathbf{p'_{i'}})$ above have a positive sign in order to obtain $\delta^3(\mathbf{x_i} - \mathbf{y_i})$ in the field anti-commutators.

The spinors, eq. 3.242, satisfy

$$\sum_{\pm s} u_{CLL}{}^+{}_\alpha(p, s)\bar{u}_{CLL}{}^+{}_\beta(p^*, s) = (2m)^{-1}[C^-R^+(i\not{p} + m)R^-C^+]_{\alpha\beta} \quad (3.245)$$

$$\sum_{\pm s} v_{CLL}{}^+{}_\alpha(p, s)\bar{v}_{CLL}{}^+{}_\beta(p^*, s) = (2m)^{-1}[C^-R^+(i\not{p} - m)R^-C^+]_{\alpha\beta}$$

where $\bar{u}_{CLL}{}^+ = u_{CLL}{}^{+\dagger}\gamma^0$ and $\bar{v}_{CLL}{}^+ = v_{CLL}{}^{+\dagger}\gamma^0$.

We now evaluate the canonical left-handed, light-front anti-commutation relation:

$$\{\psi_{CLL}{}^+{}_a(x), \psi_{CLL}{}^{+\dagger}{}_b(y)\} = \sum_{\pm s,s'} \int d^3p_i d^2p dp^+ \int d^3p_i' d^2p' dp'^+ N_{CLL}{}^+(p)\, N_{CLL}{}^+(p')\cdot$$

$$\cdot\theta(p^+)\theta(p'^+)\delta((p_i{}^3(p^+{-}p^-)/\sqrt{2} + \mathbf{p_{r\perp}}\cdot\mathbf{p_{i\perp}})/m^2)\,\delta((p_i'{}^3(p'^+ - p'^-)/\sqrt{2} + \mathbf{p'_{r\perp}}\cdot\mathbf{p'_{i\perp}})/m^2)\cdot$$

$$\cdot[\{b_{CLL}{}^{+\dagger}(p'^*,s'),b_{CLL}{}^+(p,s)\}u_{CLL}{}^+{}_a(p,s)u_{CLL}{}^{+\dagger}{}_b(p'^*,s')e^{+i(p'^*\cdot y+p'\cdot y^*)2 - i(p\cdot x+p^*\cdot x^*)/2} +$$

$$+\{d_{CLL}{}^+(p'^*,s'),d_{CLL}{}^{+\dagger}(p,s)\}v_{CLL}{}^+{}_a(p,s)v_{CLL}{}^{+\dagger}{}_b(p'^*,s')e^{-i(p'^*\cdot y+p'\cdot y^*)/2 + i(p\cdot x + p^*\cdot x^*)/2}]$$

$$= 2^{-\frac{1}{2}}\sum_{\pm s} \int d^3p_i d^2p_r dp^+ [N_{CLL}{}^+(p)]^2\theta(p^+)[\delta((p_i{}^3(p^+ - p^-)/\sqrt{2} + \mathbf{p_{r\perp}}\cdot\mathbf{p_{i\perp}})/m^2)]^2 \cdot$$

$$\cdot[u_{CLL}{}^+{}_a(p,s)u_{CLL}{}^+{}^\dagger{}_b(p^*,s)e^{+i(p^*\cdot(y-x)+p\cdot(y^*-x^*))/2} + v_{CLL}{}^+{}_a(p,s)v_{CLL}{}^{+\dagger}{}_b(p^*,s)e^{-i(p^*\cdot(y-x)+p\cdot(y^*-x^*))/2}]$$

$$= -2^{-3/2}\int d^3p_i d^2p dp^+ \theta(p^+)[N_{CLL}{}^+(p)]^2\delta'((p_i{}^3(p^+ - p^-)/\sqrt{2} + \mathbf{p_{r\perp}}\cdot\mathbf{p_{i\perp}})/m^2)(2m)^{-1}\cdot$$

$$\cdot\{[C^-R^+(i\not{p} + m)\gamma^0R^+C^-]_{ab}e^{+i(p^*\cdot(y-x)+p\cdot(y^*-x^*))/2} +$$

$$+[C^-R^+(i\not{p} - m)\gamma^0R^+C^-]_{ab}e^{-i(p^*\cdot(y-x)+p\cdot(y^*-x^*))/2}\}$$

$$= -(1/2)C^-R^+\delta_{ab}\int d^3p_i\, d^2p_\perp \int_0^\infty dp^+ \,\delta'((p_i{}^3(p^+ - p^-)/\sqrt{2} + \mathbf{p_\perp}\cdot\mathbf{p_{i\perp}})/m^2)(2\pi)^{-6}\cdot$$

$$\cdot\{e^{+i\{p^+(y^- - x^-) - \mathbf{p_{r\perp}}\cdot(\mathbf{y_{r\perp}} - \mathbf{x_{r\perp}}) + \mathbf{p_i}\cdot(\mathbf{y_i} - \mathbf{x_i})\}} + e^{-i\{p^+(y^- - x^-) - \mathbf{p_{r\perp}}\cdot(\mathbf{y_{r\perp}} - \mathbf{x_{r\perp}}) + \mathbf{p_i}\cdot(\mathbf{y_i} - \mathbf{x_i})\}}\}$$

$$= -C^-R^+\delta_{ab}(4\pi)^{-1}\int_0^\infty dp^+\delta'(\nabla_r\cdot\nabla_i/m^2)\delta^3(\mathbf{y_i}{-}\mathbf{x_i})\,\delta^2(\mathbf{y_r}{-}\mathbf{x_r})\{e^{+ip^+(y^- - x^-)} + e^{-ip^+(y^- - x^-)}\}$$

whereupon we revert back to the original form of the constraint: $\delta(\nabla_r\cdot\nabla_i/m^2)$

$$\{\psi_{CLL}{}^+{}_a(x), \psi_{CLL}{}^{+\dagger}{}_b(y)\} = -(1/2)C^-R^+\delta_{ab}\,\delta'(\nabla_r\cdot\nabla_i/m^2)\delta(y^- - x^-)\delta^2(\mathbf{y_r} - \mathbf{x_r})\delta^3(\mathbf{y_i} - \mathbf{x_i})$$

$$(3.246)$$

The result is the left-handed, light-front equivalent of the earlier non-tachyon result. Again the constraint is apparent in the anti-commutator. (The factor of 2 difference is due to light-front coordinate definitions.)

Therefore we have left-handed, light-front quantized tachyonic complexons with the equivalent of canonical anti-commutation relations, and with localized tachyonic complexons. As a result we have a canonical tachyonic complexon Quantum Field Theory.

3.5.3.2 Left-handed Case 4: Tachyonic Complexon Feynman Propagator

The light-front Feynman propagator for the left-handed ψ_{CLL}^+ tachyonic complexon field is

$$iS^+_{CLLF}(x,y) = \theta(x^+ - y^+)<0|\psi_{CLL}^+(x)\psi_{CLL}^{++}(y)\gamma^0|0> - \theta(y^+ - x^+)<0|\psi_{CLL}^{++}(y)\gamma^0\psi_{CLL}^+(x)|0>$$

$$(3.247)$$

$$= -\tfrac{1}{2}\int d^3p_i d^2p_r dp^+ \theta(p^+) N_{CLL}^{+2} \delta'((p_i^3(p^+ - p^-)/\sqrt{2} + \mathbf{p}_{r\perp}\cdot\mathbf{p}_{i\perp})/m^2)(2m)^{-1}C^-R^+\cdot$$
$$\cdot\{\theta(x^+ - y^+)[(i\not{p} + m)\gamma^0]e^{+i(p^*\cdot(y-x)+p\cdot(y^*-x^*))/2} +$$
$$+ \theta(y^+ - x^+)[(i\not{p} - m)\gamma^0]e^{-i(p^*\cdot(y-x)+p\cdot(y^*-x^*))/2}\}R^+C^-\gamma^0$$

If we define the on-shell momentum variables

$$p_0^- = (p_{r0}^1 p_{r0}^1 + p_{r0}^2 p_{r0}^2 - \mathbf{p}_{i0}\cdot\mathbf{p}_{i0} - m^2)/(2p_0^+)$$
$$p_0^+ = p^+, \quad p_{r0}^j = p_r^j \quad (\text{for } j = 1, 2),$$
$$\mathbf{p}_{i0} = \mathbf{p}_i, \quad p_{r\perp 0}^2 = p_{r0}^j p_{r0}^j$$
$$\not{p}_0 = p_0\cdot\gamma$$

with $p_0 = (p^0, \mathbf{p}_{r0} + i\mathbf{p}_{r0})$ then the above equation can be rewritten as

$$= -\tfrac{1}{2}C^-R^+\int d^4p d^3p_i N_{CLL}^{+2}\delta'((p_{i0}^3(p_0^+ - p_0^-)/\sqrt{2} + \mathbf{p}_{r\perp 0}\cdot\mathbf{p}_{i\perp 0})/m^2)(4\pi m)^{-1}e^{+i(p^*\cdot(y-x)+p\cdot(y^*-x^*))/2}\cdot$$
$$\cdot\{\theta(p^+)(i\not{p}_0 + m)\gamma^0]/[p^- - p_0^- + i\varepsilon] + \theta(-p^+)(i\not{p}_0 - m)\gamma^0]/[p^- + p_0^- - i\varepsilon]\}R^+C\gamma^0$$

$$= -\tfrac{1}{2}\int d^4p_r d^3p_i N_{CLL}^{+2}\delta'((p_{i0}^3(p^+ - p^-)/\sqrt{2} + \mathbf{p}_{r\perp}\cdot\mathbf{p}_{i\perp})/m^2)(p^+/4\pi m)e^{+i(p^*\cdot(y-x)+p\cdot(y^*-x^*))/2}\cdot$$
$$\cdot[C^-R^+(i\not{p} + m)\gamma^0 R^+C^-\gamma^0][(p^2 + m^2 + i\varepsilon)]^{-1}$$

with $p_r = (p^0, \mathbf{p}_r)$ and $p = (p^0, \mathbf{p}_r + i\mathbf{p}_r)$. Substituting for N_{CLL} and using $x\delta'(x) = -\delta(x)$ we obtain

$$= -\tfrac{1}{2}\int d^4p_r d^3p_i(2\pi)^{-7}\delta'(\mathbf{p}_r\cdot\mathbf{p}_i/m^2)\exp[ip^0(y^0 - x^0) - i\mathbf{p}_r\cdot(\mathbf{y}_r - \mathbf{x}_r) + i\mathbf{p}_i\cdot(\mathbf{y}_i - \mathbf{x}_i)]\cdot$$
$$\cdot[C^-R^+(i\not{p} + m)R^-C^+]/(p^2 + m^2 + i\varepsilon)$$

since $C^-R^+(i\not{p} + m)\gamma^0 R^+C^-\gamma^0 = C^-R^+(i\not{p} + m)R^-C^+$. The integral can be written:

$$= \int d^4p_r d^3p_i \delta'(\mathbf{p_r \cdot p_i}/m^2) C^- R^+(i\not{p} + m) R^- C^+ \cdot$$
$$\cdot \exp[-ip^0(x^0-y^0) + i\mathbf{p_r \cdot (x_r-y_r)} - i\mathbf{p_i \cdot (x_i-y_i)}]/(p^2 + m^2 + i\varepsilon)$$

$$= \int d^4p_r dM^2 \delta'(\mathbf{\nabla_r \cdot \nabla_i}/m^2) C^- R^+(ip^0\gamma^0 - (\mathbf{\nabla_r} - i\mathbf{\nabla_i})\cdot\gamma + m) R^- C^+ \cdot$$
$$\cdot \exp[-ip^0(x^0-y^0) + i\mathbf{p_r \cdot (x_r-y_r)}] J_2(\mathbf{x_i} - \mathbf{y_i}, M^2)/(p_r^2 + M^2 + i\varepsilon)$$

where

$$J_2(\mathbf{x_i} - \mathbf{y_i}, M^2) = (2\pi)^{-3}\int d^3p_i\, \delta(M^2 - \mathbf{p_i}^2 - m^2)\, \exp[-i\mathbf{p_i\cdot(x_i-y_i)}] \qquad (3.248)$$
$$= (2\pi)^{-2}|\mathbf{x_i-y_i}|^{-1}\theta(M^2 - m^2)\sin((M^2 - m^2)^{\frac{1}{2}}|\mathbf{x_i-y_i}|)$$

This tachyonic complexon Feynman propagator can be rearranged into the form of a spectral integral:

$$iS^+_{CLLF}(x, y) = -\int dM\, C^- R^+(\gamma^0\partial/\partial x^0 + (\mathbf{\nabla_r} - i\mathbf{\nabla_i})\cdot\gamma - m) R^- C^+ \delta'(\mathbf{\nabla_r \cdot \nabla_i}/m^2) \cdot$$
$$\cdot J_2(\mathbf{x_i} - \mathbf{y_i}, M^2)\Delta_{FT}(x - y, M) \qquad (3.249)$$

with $\mathbf{\nabla_r}$ and $\mathbf{\nabla_i}$ being derivatives with respect to $\mathbf{x_r}$ and $\mathbf{x_i}$ and where

$$\Delta_{FT}(x - y, M) = (2\pi)^{-4}\int d^4p_r \exp[-ip^0(x^0 - y^0) + i\mathbf{p_r \cdot (x_r - y_r)}]/(p_r^2 + M^2 + i\varepsilon) \qquad (3.250)$$

3.5.3.3 Case 5: Right-Handed Tachyonic Complexons

The case of right-handed tachyonic complexons is similar to left-handed complexons with only one difference: a minus sign in the canonical right-handed equal-time commutation relations resulting in a minus sign in the creation and annihilation operator anti-commutation relations. The right-handed tachyonic complexon wave function light-front Fourier expansion is:

$$\psi_{CLR}^+(\mathbf{x_r}, \mathbf{x_i}) = \sum_{\pm s} \int d^2p_r dp^+ d^3p_i\, N_{CLR}^+(p)\theta(p^+)\delta((p_i^3(p^+ - p^-)/\sqrt{2} + \mathbf{p_{r\perp} \cdot p_{i\perp}})/m^2) \cdot$$
$$\cdot [b_{CLR}^+(p, s)u_{CLR}^+(p, s)e^{-i(p\cdot x + p^*\cdot x^*)/2} + d_{CLR}^{+\dagger}(p, s)v_{CLR}^+(p, s)e^{+i(p\cdot x + p^*\cdot x^*)/2}] \qquad (3.251)$$

where

$$N_{CLR}^+(p) = (2\pi)^{-3}(2m/p^+)^{\frac{1}{2}} \qquad (3.252)$$

Its Hermitean conjugate is

$$\psi_{CLR}^{+\dagger}(\mathbf{x_r}, \mathbf{x_i}) = \sum_{\pm s} \int d^2p_r dp^+ d^3p_i\, N_{CLR}^+(p)\theta(p^+)\delta((p_i^3(p^+ - p^-)/\sqrt{2} + \mathbf{p_{r\perp} \cdot p_{i\perp}})/m^2) \cdot$$
$$\cdot [b_{CLR}^\dagger(p^*, s)u_{CLR}^\dagger(p^*, s)e^{+i(p^*\cdot x + p\cdot x^*)/2} + d_{CLR}(p^*, s)v_{CLR}^\dagger(p^*, s)e^{-i(p^*\cdot x + p\cdot x^*)/2}] \qquad (3.253)$$

where $\mathbf{p} = \mathbf{p_r} + i\mathbf{p_i}$, $x = x_r - ix_i$, $p\cdot x = p^0 x^0 - \mathbf{p \cdot x}$, and † indicates Hermitean conjugate.

The right-handed spinors are

$$u_{CLR}{}^{+}(p, s) = C^{+} R^{+} S_{CR}w^{1}(0)$$
$$u_{CLR}{}^{+}(p, -s) = C^{+} R^{+} S_{CR}w^{2}(0)$$
$$v_{CLR}{}^{+}(p, s) = C^{+} R^{+} S_{CR}w^{3}(0)$$
$$v_{CLR}{}^{+}(p, -s) = C^{+} R^{+} S_{CR}w^{4}(0)$$
$$u_{CLR}{}^{++}(p^*, s) = w^{1T}(0)S_{CR}R^{+}C^{+}$$
$$u_{CLR}{}^{++}(p^*, -s) = w^{2T}(0)S_{CR}R^{+}C^{+}$$
$$v_{CLR}{}^{++}(p^*, s) = w^{3T}(0)S_{CR}R^{+}C^{+}$$
$$v_{CLR}{}^{++}(p^*, -s) = w^{4T}(0)S_{CR}R^{+}C^{+}$$

$$(3.254)$$

where the superscript "T" indicates the transpose. The anti-commutation relations of the Fourier coefficient operators are

$$\{b_{CLR}(p,s), b_{CLR}{}^{\dagger}(p'^*,s')\} = -2^{-\frac{1}{2}}\delta_{ss'}\delta(p^{+} - p'^{+})\delta^{2}(\mathbf{p_r} - \mathbf{p'_{r'}})\delta^{3}(\mathbf{p_i} + \mathbf{p'_{i'}})$$
$$\{d_{CLR}(p,s), d_{CLR}{}^{\dagger}(p'^*,s')\} = -2^{-\frac{1}{2}}\delta_{ss'}\cdot\delta(p^{+} - p'^{+})\delta^{2}(\mathbf{p_r} - \mathbf{p'_{r'}})\delta^{3}(\mathbf{p_i} + \mathbf{p'_{i'}})$$
$$\{b_{CLR}(p,s), b_{CLR}(p'^*,s')\} = \{d_{CLR}(p,s), d_{CLR}(p'^*,s')\} = 0$$
$$\{b_{CLR}{}^{\dagger}(p,s), b_{CLR}{}^{\dagger}(p'^*,s')\} = \{d_{CLR}{}^{\dagger}(p,s), d_{CLR}{}^{\dagger}(p'^*,s')\} = 0$$
$$\{b_{CLR}(p,s), d_{CLR}{}^{\dagger}(p'^*,s')\} = \{d_{CLR}(p,s), b_{CLR}{}^{\dagger}(p'^*,s')\} = 0$$
$$\{b_{CLR}{}^{\dagger}(p,s), d_{CLR}{}^{\dagger}(p'^*,s')\} = \{d_{CLR}(p,s), b_{CRR}(p'^*,s')\} = 0$$

$$(3.255)$$

The spinors satisfy

$$\sum_{\pm s} u_{CLR}{}^{+}{}_{\alpha}(p, s)\bar{u}_{CLR}{}^{+}{}_{\beta}(p^*, s) = (2m)^{-1}[C^{+}R^{+}(-i\not{p} + m)R^{-}C^{-}]_{\alpha\beta} \qquad (3.256)$$

$$\sum_{\pm s} v_{CLR}{}^{+}{}_{\alpha}(p, s)\bar{v}_{CLR}{}^{+}{}_{\beta}(p^*, s) = (2m)^{-1}[C^{+}R^{+}(-i\not{p} - m)R^{-}C^{-}]_{\alpha\beta}$$

where $\bar{u}_{CLR}{}^{+} = u_{CLR}{}^{++\dagger}\gamma^{0}$ and $\bar{v}_{CLR}{}^{+} = v_{CLR}{}^{++\dagger}\gamma^{0}$.

The right-handed anti-commutation relation with a minus sign follows in particular because of the minus signs in eqs. 3.255.

3.5.3.4 Right-handed Case 5: Tachyonic Complexon Feynman Propagator

The Feynman propagator for right-handed tachyonic complexons can be obtained from eqs. 3.249 and 3.250 by changing the parity projection operator and some numerator signs in the integral (basically setting $p \rightarrow -p$) resulting in

$$iS^{+}{}_{CLRF}(x, y) = \int dM\, C^{+}R^{+}(\gamma^{0}\partial/\partial x^{0} + (\nabla_r - i\nabla_i)\cdot\gamma - m)R^{-}C^{-}\,\delta'(\nabla_r\cdot\nabla_i/m^{2})\,\cdot$$
$$\cdot J_{2}(\mathbf{x_i} - \mathbf{y_i}, M^{2})\triangle_{FT}(x - y, M) \qquad (3.257)$$

with $\nabla_r + i\nabla_i$ derivatives with respect to $\mathbf{x}_r$ and $\mathbf{x}_i$ and where

$$\Delta_{FT}(x - y, M) = (2\pi)^{-4}\!\int\! d^4p_r \, \exp[-ip^0(x^0 - y^0) + i\mathbf{p}_r\cdot(\mathbf{x}_r - \mathbf{y}_r)]/(p_r^{\,2} + M^2 + i\varepsilon)$$

$$(3.258)$$

3.5.3.5 Other Cases? No

The four cases considered above are the only cases having symmetry under the real Lorentz group L and a single real energy (with a corresponding single real time parameter) that is independent of the direction of the boost thus preserving (real) spatial rotation invariance. The reality of the time variable survives the breakdown to conventional Lorentz invariance.

One might think that using the other type of spinor boost operator.

$$S_{CR}(\Lambda_{CR}(\omega, \hat{\mathbf{w}})) = \exp(-i\omega_R\sigma_{0i}w_i/2) = \exp(-\omega_R\gamma^0\boldsymbol{\gamma}\cdot\hat{\mathbf{w}}/2) \qquad (3.259)$$
$$= \cosh(\omega_R/2)I + \sinh(\omega_R/2)\gamma^0\boldsymbol{\gamma}\cdot\hat{\mathbf{w}}$$

where $\omega_R = \omega - i\pi/2$ might lead to more possible forms of spin ½ wave equations and particles. In fact it merely leads to the same particle types but with the role of the left-handed and right-handed fields reversed. The result would be a "right-handed" Standard Model contrary to experiment.

3.6 Spinor Boosts Generate 4 Species of Particles: Leptons and Quarks

In this chapter we have found four types of fermions using complex Lorentz boosts that correspond in a natural way with the four general *species* (types) of known fermions: charged leptons, neutrinos, up-type color quarks and down-type color quarks.[27]

3.6.1 Charged lepton fermions

The conventional Dirac equation and solutions.

3.6.2 Neutrinos

Simple tachyons with real energy and 3-momentum. Their free field equation is:

$$(\gamma^\mu\partial/\partial x^\mu - m)\psi_T(x) = 0 \qquad (3.260)$$

and their left-handed $\psi_{TL}{}^+$ Feynman propagator is:

$$iS^+{}_{TLF}(x, y) = \tfrac{1}{2}C^-R^+\gamma^0\!\int\! d^4p(2\pi)^{-4}\, p^+e^{-ip\cdot(x-y)}/(p^2 + m^2 + i\varepsilon) \qquad (3.261)$$

Similarly the light-front Feynman propagator for the right-handed $\psi_{TR}{}^+$ tachyon field is

---

[27] We call each type of fermion a *species*. Each species has three known generations.

$$iS^+{}_{TRF}(x,y) = -\tfrac{1}{2}C^+R^+\gamma^0\!\int d^4p(2\pi)^{-4}\, p^+e^{-ip\cdot(x-y)}/(p^2 + m^2 + i\varepsilon) \qquad (3.262)$$

3.6.3 Up-type Color Quarks

Up-type quarks are assumed[28] to be fermions with complex 3-momenta - complexons, and an internal color SU(3) symmetry, that satisfy $p^2 = m^2$. Their field equation with a color SU(3) index, denoted a, inserted is

$$[i\gamma^0\partial/\partial t + i\gamma\cdot(\nabla_r + i\nabla_i) - m]\psi_C{}^a(t,\, \mathbf{x_r},\, \mathbf{x_i}) = 0 \qquad (3.263)$$

with the subsidiary condition

$$\nabla_r\cdot\nabla_i\, \psi_C{}^a(t,\, \mathbf{x_r},\, \mathbf{x_i}) = 0 \qquad (3.264)$$

The free field solution is:

$$\psi_C{}^a(x) = \sum_{\pm s}\int d^3p_r d^3p_i\, N_C(p)\delta(\mathbf{p_r}\cdot\mathbf{p_i}/m^2)[b_C(p,a,s)u_C{}^a(p,\, s)e^{-i(p\cdot x + p^*\cdot x^*)/2} +$$
$$+\, d_C{}^\dagger(p,a,s)v_C{}^a(p,\, s)e^{+i(p\cdot x + p^*\cdot x^*)/2}] \qquad (3.265)$$

The free Feynman propagator arranged into the form of a spectral integral is

$$iS_C{}^{ab}(x,y)= -\delta^{ab}\!\int dM\, (i\gamma^0\partial/\partial x^0 - i(\nabla_r - i\nabla_i)\cdot\gamma + m)\delta'(\nabla_r\cdot\nabla_i/m^2)J(\mathbf{x_i} - \mathbf{y_i},\, M^2)\Delta_F(x - y,\, M)$$
$$(3.266)$$

where

$$\Delta_F(x - y,\, M) = (2\pi)^{-4}\!\int d^4p_r\, \exp[-ip^0(x^0 - y^0) + i\mathbf{p_r}\cdot(\mathbf{x_r} - \mathbf{y_r})]/(p_r{}^2 - M^2 + i\varepsilon) \qquad (3.267)$$

and

$$J(\mathbf{x_i},\, M^2) = (2\pi)^{-3}\!\int d^3p_i\, \delta(M^2 + \mathbf{p_i}{}^2 - m^2)\, \exp[-i\mathbf{p_i}\cdot(\mathbf{x_i} - \mathbf{y_i})] \qquad (3.268)$$
$$= (2\pi)^{-2}|\mathbf{x_i} - \mathbf{y_i}|^{-1}\theta(m^2 - M^2)\sin((m^2 - M^2)^{1/2}|\mathbf{x_i} - \mathbf{y_i}|)$$

3.6.4 Down-type Color Quarks

Tachyonic complexons with complex 3-momenta, and an internal global SU(3) symmetry, that have mass shell condition $p^2 = -m^2$. Their field equation with a color SU(3) index, denoted a, inserted is

$$[\gamma^0\partial/\partial t + \gamma\cdot(\nabla_r + i\nabla_i) - m]\psi_{CL}{}^a(t,\, \mathbf{x_r},\, \mathbf{x_i}) = 0 \qquad (3.269)$$

with the subsidiary condition on the wave function

---

[28] The complexon theory that we develop and use for quark dynamics in the Standard Model is <u>not</u> required. Our Standard Model could use Dirac fermion dynamics for the up-type quarks and tachyon dynamics for down-type quarks. Then the (broken) Left-handed complex Lorentz boosts would have the basic space-time group rather than L_C. We choose to use complexon dynamics for quarks because they have an internal SU(3)-like structure suggestive of color SU(3). More importantly, their spin dynamics is different and thus may resolve the differences between theory and experiment for the deep inelastic parton spin-dependent structure functions.

$$\nabla_r \cdot \nabla_i \, \psi_{CL}{}^a(t, \mathbf{x_r}, \mathbf{x_i}) = 0 \tag{3.270}$$

Its free field left-handed solution is:

$$\psi_{CLL}{}^{+a}(\mathbf{x_r}, \mathbf{x_i}) = \sum_{\pm s} \int d^2 p_r dp^+ d^3 p_i \, N_{CLL}{}^+(p)\theta(p^+)\delta((p_i{}^3(p^+ - p^-)/\surd2 + \mathbf{p_{r\perp}} \cdot \mathbf{p_{i\perp}})/m^2) \cdot$$
$$\cdot [b_{CLL}{}^+(p,a,s)u_{CLL}{}^a(p,a,s)e^{-i(p\cdot x + p^*\cdot x^*)/2} + d_{CLL}{}^{++}(p,a,s)v_{CLL}{}^{+a}(p,a,s)e^{+i(p\cdot x + p^*\cdot x^*)/2}] \tag{3.271}$$

and its right-handed solution is

$$\psi_{CLR}{}^{+a}(\mathbf{x_r}, \mathbf{x_i}) = \sum_{\pm s} \int d^2 p_r dp^+ d^3 p_i \, N_{CLR}{}^+(p)\theta(p^+)\delta((p_i{}^3(p^+ - p^-)/\surd2 + \mathbf{p_{r\perp}} \cdot \mathbf{p_{i\perp}})/m^2) \cdot$$
$$\cdot [b_{CLR}{}^+(p,a,s)u_{CLR}{}^{+a}(p,a,s)e^{-i(p\cdot x + p^*\cdot x^*)/2} + d_{CLR}{}^{++}(p,a,s)v_{CLR}{}^{+a}(p,a,s)e^{+i(p\cdot x + p^*\cdot x^*)/2}] \tag{3.272}$$

The free left-handed Feynman propagator arranged into the form of a spectral integral is

$$iS^+_{CLLF}{}^{ab}(x,y) = -\delta^{ab}\int dM \, C^-R^+(\gamma^0\partial/\partial x^0 + (\nabla_r - i\nabla_i)\cdot\gamma - m)R^-C^+\delta'(\nabla_r\cdot\nabla_i/m^2) \cdot$$
$$\cdot J_2(\mathbf{x_i} - \mathbf{y_i}, M^2)\Delta_{FT}(x - y, M) \tag{3.273}$$

with ∇_r and ∇_i derivatives with respect to $\mathbf{x_r}$ and $\mathbf{x_i}$ and where

$$\Delta_{FT}(x - y, M) = (2\pi)^{-4}\int d^4 p_r \exp[-ip^0(x^0 - y^0) + i\mathbf{p_r}\cdot(\mathbf{x_r} - \mathbf{y_r})]/(p_r{}^2 + M^2 + i\varepsilon) \tag{3.274}$$

and

$$J_2(\mathbf{x_i}, M^2) = (2\pi)^{-3}\int d^3 p_i \, \delta(M^2 - \mathbf{p_i}{}^2 - m^2) \exp[-i\mathbf{p_i}\cdot(\mathbf{x_i} - \mathbf{y_i})] \tag{3.275}$$

$$= (2\pi)^{-2}|\mathbf{x_i} - \mathbf{y_i}|^{-1}\theta(M^2 - m^2)\sin((M^2 - m^2)^{\frac12}|\mathbf{x_i} - \mathbf{y_i}|)$$

The free right-handed Feynman propagator arranged into the form of a spectral integral is

$$iS^+_{CLRF}{}^{ab}(x, y) = \delta^{ab}\int dM \, C^+R^+(\gamma^0\partial/\partial x^0 + (\nabla_r - i\nabla_i)\cdot\gamma - m)R^-C^-\delta'(\nabla_r\cdot\nabla_i/m^2) \cdot$$
$$\cdot J_2(\mathbf{x_i} - \mathbf{y_i}, M^2)\Delta_{FT}(x - y, M) \tag{3.276}$$

with ∇_r and ∇_i derivatives with respect to $\mathbf{x_r}$ and $\mathbf{x_i}$, and where

$$\Delta_{FT}(x - y, M) = (2\pi)^{-4}\int d^4 p_r \exp[-ip^0(x^0 - y^0) + i\mathbf{p_r}\cdot(\mathbf{x_r} - \mathbf{y_r})]/(p_r{}^2 + M^2 + i\varepsilon) \tag{3.277}$$

Appendix 3-A. Leptonic Tachyon Spinors

The general form of the solutions of the free tachyon Dirac equation can be written

$$\psi_T^{\,r}(x) = e^{-i\chi_r\, p\cdot x}\, w^r(p) \tag{3-A.1}$$

where $\chi_r = +1$ for $r = 1, 2$ and $\chi_r = -1$ for $r = 3, 4$. Denoting the spinors $w^r(p) = w^r(0)$ for a particle is at rest in a frame $(E = m)$ we see they can take the form

$$w^r(0) \;=\; \begin{bmatrix} \delta_{1r} \\ \delta_{2r} \\ \delta_{3r} \\ \delta_{4r} \end{bmatrix} \tag{3-A.2}$$

where Kronecker deltas appear in the brackets. From eq. 3.30 we find

$$S_L(\Lambda_L(\omega, \mathbf{u}))w^r(0) = w_T^{\,r}(p) \tag{3-A.3}$$

Using eq. 3.66 for $S_L(\Lambda_L(\omega, \mathbf{u}))$ and

$$\mathbf{p} = m\mathbf{v}\gamma_s \qquad\qquad E = m\gamma_s \tag{3-A.4}$$

we see that eq. 3-A.3 implies the columns of the resulting $S_L(\Lambda_L(\omega, \mathbf{u}))$ matrix are

$$S_L(\Lambda_L(\omega, \mathbf{u})) = \begin{array}{cccc} \underline{w_T^{\,3}(p)} & \underline{w_T^{\,4}(p)} & \underline{w_T^{\,1}(p)} & \underline{w_T^{\,2}(p)} \\[4pt] \left[\begin{array}{cccc} \cosh(\omega_L/2) & 0 & \sinh(\omega_L/2)p_z/p & \sinh(\omega_L/2)p_-/p \\[6pt] 0 & \cosh(\omega_L/2) & \sinh(\omega_L/2)p_+/p & -\sinh(\omega_L/2)p_z/p \\[6pt] \sinh(\omega_L/2)p_z/p & \sinh(\omega_L/2)p_-/p & \cosh(\omega_L/2) & 0 \\[6pt] \sinh(\omega_L/2)p_+/p & -\sinh(\omega_L/2)p_z/p & 0 & \cosh(\omega_L/2) \end{array}\right] \end{array}$$

$$\tag{3-A.5}$$

based on the superluminal transformation of positive energy states to negative energy states with $p_\pm = p_x \pm ip_y$ and where $p = |\mathbf{p}|$. It is easy to verify

$$(i\not{p} - \chi_r m)w_T^{\,r}(p) = 0 \qquad\qquad (3\text{-A}.6)$$

where $\chi_r = -1$ for $r = 1, 2$ and $\chi_r = +1$ for $r = 3, 4$.

The spinors that we defined earlier can be generalized in a manner similar to Dirac spinors. We will use a similar notation to the Dirac spinor notation:

$$\begin{aligned}
u_T(p, s) &= w_T^{\,1}(p) \\
u_T(p, -s) &= w_T^{\,2}(p) \\
v_T(p, s) &= w_T^{\,3}(p) \\
v_T(p, -s) &= w_T^{\,4}(p)
\end{aligned} \qquad\qquad (3\text{-A}.7)$$

We define "double dagger" spinors:

$$\begin{aligned}
u_T^{\ddagger}(p, s) &= u_T^{\dagger}(p, s)i\boldsymbol{\gamma}\cdot\mathbf{p}/|\mathbf{p}| \\
u_T^{\ddagger}(p, -s) &= u_T^{\dagger}(p, -s)i\boldsymbol{\gamma}\cdot\mathbf{p}/|\mathbf{p}| \\
v_T^{\ddagger}(p, s) &= v_T^{\dagger}(p, s)i\boldsymbol{\gamma}\cdot\mathbf{p}/|\mathbf{p}| \\
v_T^{\ddagger}(p, -s) &= v_T^{\dagger}(p, -s)i\boldsymbol{\gamma}\cdot\mathbf{p}/|\mathbf{p}|
\end{aligned} \qquad\qquad (3\text{-A}.8)$$

where † indicates Hermitean conjugate, which appear in important spinor "completeness" sums:

$$\sum_{\pm s} u_{T\alpha}(p, s)u_T^{\ddagger}{}_\beta(p, s) = (2m)^{-1}(i\not{p} - m)_{\alpha\beta} \qquad\qquad (3\text{-A}.9)$$

$$\sum_{\pm s} v_{T\alpha}(p, s)v_T^{\ddagger}{}_\beta(p, s) = (2m)^{-1}(i\not{p} + m)_{\alpha\beta} \qquad\qquad (3\text{-A}.10)$$

or

$$\sum_{\pm s} u_{T\alpha}(p, s)u_T^{\dagger}{}_\beta(p, s) = -i(2m)^{-1}[(i\not{p} - m)\boldsymbol{\gamma}\cdot\mathbf{p}/|\mathbf{p}|]_{\alpha\beta} \qquad\qquad (3\text{-A}.11)$$

$$\sum_{\pm s} v_{T\alpha}(p, s)v_T^{\dagger}{}_\beta(p, s) = -i(2m)^{-1}[(i\not{p} + m)\boldsymbol{\gamma}\cdot\mathbf{p}/|\mathbf{p}|]_{\alpha\beta} \qquad\qquad (3\text{-A}.12)$$

Lastly we define light-front, left-handed tachyon spinors by

$$\begin{aligned}
u_{TL}^{\;+}(p, s) &= C^- R^+ S_L(\Lambda_L(\omega, \mathbf{u}))w^1(0) \\
u_{TL}^{\;+}(p, -s) &= C^- R^+ S_L(\Lambda_L(\omega, \mathbf{u}))w^2(0) \\
v_{TL}^{\;+}(p, s) &= C^- R^+ S_L(\Lambda_L(\omega, \mathbf{u}))w^3(0) \\
v_{TL}^{\;+}(p, -s) &= C^- R^+ S_L(\Lambda_L(\omega, \mathbf{u}))w^4(0)
\end{aligned} \qquad\qquad (3\text{-A}.13)$$

$$\begin{aligned}
u_{TL}^{\;++}(p, s) &= w^{1T}(0)\, S_L^{\dagger}(\Lambda_L(\omega, \mathbf{u}))\, R^+ C^- \\
u_{TL}^{\;++}(p, -s) &= w^{2T}(0)\, S_L^{\dagger}(\Lambda_L(\omega, \mathbf{u}))R^+ C^- \\
v_{TL}^{\;++}(p, s) &= w^{3T}(0)\, S_L^{\dagger}(\Lambda_L(\omega, \mathbf{u}))R^+ C^- \\
v_{TL}^{\;++}(p, -s) &= w^{4T}(0)\, S_L^{\dagger}(\Lambda_L(\omega, \mathbf{u}))R^+ C^-
\end{aligned} \qquad\qquad (3\text{-A}.14)$$

where the superscript "T" indicates the transpose and † indicates Hermitean conjugate.

4. Bradyon PseudoFermion Quantum Fields

This chapter develops a two coordinate system formulation for a two bradyon fermion that for what we call *duplex fermions*.[29] The following chapter extends the formalism to the case of a particle with a bradyon external coordinate system and a tachyon internal coordinate system. The goal is to develop quantum field theories for both the external and internal features of elementary particles. Our past efforts led to the gambol formulation of the interiors of elementary particles. We now look towards the formulation of fermion quantum fields embodying both the Lorentz group $SO^+(1, 3)$ and the ElectroWeak $SU(2) \otimes U(1)$ group.

This chapter uses *PseudoFermion*[30] quantum fields.

4.1 PseudoFermions

We define[31] a PseudoFermion, with the form

$$\psi_{i\alpha\beta}(y, z) \tag{4.1}$$

where y and z are independent bradyon coordinates in r space-time dimensions, where i =1, 2 labels the PseudoQuantum fields, and α and β are spinor indices.[32]

We begin by defining a free PseudoFermion PseudoQuantum Lagrangian with two related quantum fields ψ_1 and ψ_2 that are functions of two sets of bradyon coordinates in r space-time dimensions, y and z,

$$\mathscr{L} = \overline{\psi}_{2BB}{}^{\alpha\beta}[-M^{-1}\gamma_{y\alpha\kappa}{}^{\mu}\partial/\partial y^{\mu}\gamma_{z\beta\lambda}{}^{\nu}\partial/\partial z^{\nu} - M]\psi_{1BB}{}^{\kappa\lambda} +$$
$$+ \overline{\psi}_{1BB}{}^{\alpha\beta}[-M^{-1}\gamma_{y\alpha\kappa}{}^{\mu}\partial/\partial y^{\mu}\gamma_{z\beta\lambda}{}^{\nu}\partial/\partial z^{\nu} - M]\psi_{2BB}{}^{\kappa\lambda} \tag{4.2}$$

where $\gamma_y{}^{\mu}$ and $\gamma_z{}^{\mu}$ are Dirac matrices for the y and z coordinates respectively, and y and z are coordinates in r dimension space-times, M is the mass, and

---

[29] There are four types of duplex fermions. Each type of fermion has two parts. This chapter describes duplex fermions with two bradyon parts. The following chapters describe duplex fermions with a bradyon part and a tachyon part, with a tachyon part and a bradyon part, and with two tachyon parts.

[30] Chapter 6 of Blaha (2022c) discusses wave functions with multiple coordinates. See also (2023c).

[31] We follow the conventions of Bjorken (1965) with the $g^{\mu\nu}$ metric (1, -1, -1, -1).

[32] Appendix C of Blaha (2016f) presents a first quantized PseudoQuantum theory CQ Mechanics that embodies both classical and quantum theory. **This theory is the non-relativistic quantum mechanics limit of the PseudoFermion theory developed here.** CQ Mechanics has two sets of coordinates that combine to create a generalization of conventional Quantum Mechanics. It has applications in a generalized Feynman path integral formalism, a generalized Schrödinger equation, a generalized Boltzmann equation, the Fokker-Planck equation, a generalized approach to quantum and classical chaos, and to quantum entanglement as well as semi-quantum entanglement. Our "Pseudo" formalisms apply to both Quantum Field Theory and Quantum Mechanics. In these applications there is a clear almost continuous transition between the quantum and the classical sectors.

$$\overline{\psi}_{iBB\alpha\beta} = \psi_{i\kappa\lambda}{}^{\dagger}\, \gamma_y{}^0{}_{\kappa\alpha}\, \gamma_z{}^0{}_{\lambda\beta} \tag{4.2a}$$

for $i = 1, 2$ where the subscripts y and z indicate Dirac spinors associated with the y and z coordinates.

The equations of motion are

$$[-M^{-1}\gamma_y{}^{\mu}\partial/\partial y^{\mu}\, \gamma_z{}^{\nu}\partial/\partial z^{\nu} - M]\psi_{1BB} = 0 \tag{4.2a}$$
$$[-M^{-1}\gamma_y{}^{\mu}\cdot\partial/\partial y^{\mu}\, \gamma_z{}^{\nu}\partial/\partial z^{\nu} - M]\psi_{2BB} = 0 \tag{4.2b}$$

We define subsidiary equations of motion

$$[i\gamma_y{}^{\mu}\partial/\partial y^{\mu} - M]\psi_{jBB} = 0 \tag{4.2c}$$
$$[i\gamma_z{}^{\nu}\partial/\partial z^{\nu} - M]\psi_{jBB} = 0 \tag{4.2d}$$

for $j = 1, 2$. Eqs. 4.2c and 4.2d imply eqs. 4.2a and 4.2b.

One conjugate momentum (with two indices α, β) is

$$\pi_{y1BB\alpha\beta} = \partial\mathcal{L}/\partial(\partial\psi_{1BB\alpha\beta}/\partial y^0) = M^{-1}(\partial/\partial z^{\nu}\, \psi_2{}^{\dagger}\gamma_z{}^0\gamma_z{}^{\nu})_{\alpha\beta} = -i(\psi_2{}^{\dagger}\gamma_z{}^0)_{\alpha\beta} \tag{4.3}$$

after partial integrations (with surface terms having the value zero) of

$$L = \int d^r y \int d^r z\, \mathcal{L}$$

using the subsidiary equation of motion:

$$i\gamma_z{}^{\nu}\partial/\partial z^{\nu}\, \psi_{2BB}{}^{\dagger} = M\psi_{2BB}{}^{\dagger} \tag{4.4}$$

Similarly the other conjugate momenta are

$$\pi_{z1BB\alpha\beta} = \partial\mathcal{L}/\partial(\partial\psi_{1BB\alpha\beta}/\partial z^0) = M^{-1}(\partial/\partial y^{\nu}\psi_{2BB}{}^{\dagger}\gamma_y{}^0\gamma_y{}^{\nu})_{\alpha\beta} = -i(\psi_{2BB}{}^{\dagger}\gamma_y{}^0)_{\alpha\beta} \tag{4.5}$$
$$\pi_{y2BB\alpha\beta} = \partial\mathcal{L}/\partial(\partial\psi_{2BB\alpha\beta}/\partial y^0) = M^{-1}(\partial/\partial z^{\nu}\, \psi_{1BB}{}^{\dagger}\gamma_z{}^0\gamma_z{}^{\nu})_{\alpha\beta} = -i(\psi_{1BB}{}^{\dagger}\gamma_z{}^0)_{\alpha\beta} \tag{4.6}$$
$$\pi_{z2BB\alpha\beta} = \partial\mathcal{L}/\partial(\partial\psi_{2BB\alpha\beta}/\partial z^0) = M^{-1}(\partial/\partial y^{\nu}\, \psi_{1BB}{}^{\dagger}\gamma_y{}^0\gamma_z{}^{\nu})_{\alpha\beta} = -i(\psi_{1BB}{}^{\dagger}\gamma_y{}^0)_{\alpha\beta} \tag{4.7}$$

using the subsidiary equations of motion:

$$i\gamma_z{}^{\nu}\partial/\partial z^{\nu}\, \psi_{1BB}{}^{\dagger} = M\psi_{1BB}{}^{\dagger} \tag{4.8}$$
$$i\gamma_y{}^{\nu}\partial/\partial y^{\nu}\, \psi_{1BB}{}^{\dagger} = M\psi_{1BB}{}^{\dagger} \tag{4.9}$$
$$i\gamma_y{}^{\nu}\partial/\partial y^{\nu}\, \psi_{2BB}{}^{\dagger} = M\psi_{2BB}{}^{\dagger} \tag{4.10}$$

Note: we omit the BB subscript in the remainder of this chapter for convenience.
We define new momenta to have a $y - z$ symmetry:

$$\pi_{1\alpha\beta} = (\pi_{y1}\gamma_z{}^0)_{\alpha\beta} = (\pi_{z1}\gamma_y{}^0)_{\alpha\beta} = \psi_2{}^\dagger{}_{\alpha\beta} \tag{4.11}$$

$$\pi_{2\alpha\beta} = (\pi_{y2}\gamma_z{}^0)_{\alpha\beta} = (\pi_{z2}\gamma_y{}^0)_{\alpha\beta} = \psi_1{}^\dagger{}_{\alpha\beta} \tag{4.12}$$

The form of the conjugate momenta implies the only non-zero equal time anticommutators are:[33]

$$\{\pi_{j\alpha\beta}(\mathbf{y}, y^0, \mathbf{z}, z^0),\ \psi_{i\kappa\lambda}(\mathbf{y'}, y^0, \mathbf{z'}, z^{0'})\} = \{\ \psi_{j\alpha\beta}{}^\dagger(\mathbf{y}, y^0, \mathbf{z}, z^0),\ \psi_{i\kappa\lambda}(\mathbf{y'}, y^0, \mathbf{z'}, z^0)\}$$

$$= (1 - \delta_{ij})\ \delta_{\alpha\kappa}\delta_{\beta\lambda}\ \delta^{r-1}(\mathbf{y} - \mathbf{y'})\delta^{r-1}(\mathbf{z} - \mathbf{z'}) \tag{4.13}$$

for i, j = 1, 2.

4.1.1 Bradyon Interior Quantum Field

The free bradyon PseudoFermion wave function has the form:

$$\psi_{i\alpha\beta}(y, z) = \underset{s_1\ s_2}{\Sigma\ \Sigma} \int dp^{r-1}\int dq^{r-1}\ N(p, q)\ [b_i(p, q, s_1, s_2)u_\alpha(p,s_1)u_\beta(q, s_2)\exp(-ip\cdot y - iq\cdot z) +$$

$$+ d_i{}^\dagger(p, q, s1, s2)v_\alpha(p,s1)v_\beta(q, s2)\exp(ip\cdot y + iq\cdot z)] \tag{4.14}$$

plus Hermitean conjugates for i = 1, 2 where N(p, q) is a normalization factor. Note both parts (y and z) of ψ_i are on the mass shell: $p^2 = M^2$ and $q^2 = M^2$.

The creation and annihilation operators satisfy the anticommutation relations:

$$\{b_i(p, q, s_1, s_2), b_j{}^\dagger(p', q', s_1', s_2')\} = (1 - \delta_{ij})\ \delta_{s_1, s_1'}\ \delta_{s_2, s_2'}\delta^{r-1}(\mathbf{p} - \mathbf{p'})\delta^{r-1}(\mathbf{q} - \mathbf{q'}) \tag{4.15}$$

$$\{d_i(p, q, s_1, s_2), d_j(p', q', s_1', s_2')\} = (1 - \delta_{ij})\ \delta_{s_1, s_1'}\ \delta_{s_2, s_2'}\delta^{r-1}(\mathbf{p} - \mathbf{p'})\delta^{r-1}(\mathbf{q} - \mathbf{q'}) \tag{4.16}$$

The other anticommutators have zero values.

They lead to the equal time anticommutation relations of eq. 4.13. The result is the free external and internal bradyon PseudoFermion formalism.

4.1.2 Bradyon Interior Gamboled[34] Quantum Field

If we introduce fractionation s where the particle is fractionated to s gambols, and internal symmetry index γ, the free PseudoFermion wave function has the form:

$$\psi_{i\alpha\beta\gamma}(y, z) = \underset{s\ s_1\ s_2}{\Sigma\Sigma\Sigma} \int dp^{r-1}\int dq^{r-1}\ N(p, q)[b_{\gamma i}(s, p, q, s_1, s_2)u_\alpha(p,s_1)u_\beta(q,s_2)\exp(-ip\cdot y - iq\cdot z) +$$

$$+ d_{\gamma i}(s, p, q, s_1, s_2)^\dagger v_\alpha(p,s_1)v_\beta(q,s_2)\exp(ip\cdot y + iq\cdot z)] \tag{4.17}$$

plus Hermitean conjugates for i = 1, 2 where s_i is the spin and N(p, q) is a normalization factor.

The composite creation and annihilation operators satisfy the anti-commutation relations:

---

[33] See S. Blaha, Il Nuovo Cimento **49A**, 35 (1979) for one coordinate system, PseudoQuantum fermions.

[34] We will use gambol interchangeably with interior in the remainder of this chapter.

$$\{b_{\gamma i}(s, p, q, s_1, s_2), b_{\gamma j}(s', p', q', s_1', s_2')^{\dagger}\} = \{d_{\gamma i}(s, p, q, s_1, s_2), d_{\gamma j}(s', p', q', s_1', s_2')^{\dagger}\} =$$
$$= (1 - \delta_{ij}) \, \delta_{s,s'} \, \delta_{s_1,s_1'} \, \delta_{s_2,s_2'} \delta^{r-1}(\mathbf{p} - \mathbf{p'}) \delta^{r-1}(\mathbf{q} - \mathbf{q'}) \, U_g(\varepsilon(s,q)) \qquad (4.18)$$

where U_g is given by the gambol *Planckian* distribution *for particles*:[35]

$$U_g(\varepsilon(s, q, k)) = 15 \, N \, (\pi kT)^{-4} \, \varepsilon^3 / (e^{\varepsilon/kT} - 1) \qquad (4.19)$$

where N is a normalization constant and

$$\varepsilon(s. \, q, \, k) = [(m_g s + M)/(s + 1)]q \cdot k / M \qquad (4.20)$$

where s is the fractionation parameter, and m_g is the gambol mass. The non-zero anti-commutators are

$$\{b_{\gamma i}(s, p, q, s_1, s_2), b_{\gamma j}(s', p', q', s_1', s_2')\} = 0 \qquad (4.21)$$
$$\{b_{\gamma i}(s, p, q, s_1, s_2)^{\dagger}, b_{\gamma j}(s', p', q', s_1', s_2')^{\dagger}\} = 0$$
$$\{d_{\gamma i}(s, p, q, s_1, s_2), d_{\gamma j}(s', p', q', s_1', s_2')\} = 0$$
$$\{d_{\gamma i}(s, p, q, s_1, s_2)^{\dagger}, d_{\gamma j}(s', p', q', s_1', s_2')^{\dagger}\} = 0$$

The other composite anti-commutation operators and the anti-commutators of b and d type operators are all zero. They lead to the equal time anticommutation relations of eq. 4.13. The result is the free gamboled bradyon PseudoFermion formalism. PseudoBoson quantum fields may be similarly defined.[36]

4.1.3 Extraction of Gambol and Particle Operators

The creation/annihilation operators may be extracted from the wave function of eq. 4.17 by integrating over the $(r - 1)$ – spatial coordinates:

$$\int dz^{r-1} e^{ik \cdot z} \psi_{i\alpha\beta\gamma}(y,z) = (2\pi)^{r-1} \{ \sum_{s} \sum_{s_1} \sum_{s_2} \int dp^{r-1} N(p,k)[b_{\gamma i}(s,p,k,s_1, s_2)u_\alpha(p,s_1)u_\beta(k,s_2) \, e^{-ip \cdot y} +$$

$$+ d_{\gamma i}(s, p, -k, s_1, s_2)^{\dagger} v_\alpha(p,s_1)v_\beta(-k,s_2) \, e^{ip \cdot y}]\}$$

where $-k = (k^0, -\mathbf{k})$. This integration and similar integrations lead to the form of particle and gambol creation/annihilation operators seen below.

4.2 Interior Gambol Creation/Annihilation Operators

We define gambol operators with

$$b^{1/s}_{g\gamma i}(s, q, s_2) = \sum_{s_1} (\int d^{r-1}p N'(p,q))^{\frac{1}{2}} \, b^{1/s}_{\gamma i}(s, p, q, s_1, s_2) \qquad (4.22)$$

---

[35] Blaha (2024b).

[36] See Blaha (2022c) for the description of pseudoboson quantum fields: free scalar pseudobosons, Free U(1) Vector PseudoBosons, Non-Abelian Vector PseudoBosons, and PseudoGraviton PseudoQuantum Gravity.

$$b^{1/s}{}_{g\gamma i}(s, q, s_2)^{\dagger} = \Sigma_{s_1} (\int d^{r-1}p N'(p,q))^{\frac{1}{2}} b^{1/s}{}_{\gamma i}(s, p, q, s_1, s_2)^{\dagger}$$

where s is the fractionation and $N'(p,q) = (2\pi)^{r-1}N(p,q)$. Then we define the gambol field operator anti-commutators with the anti-commutation relations:

$$\{b^{1/s}{}_{g\gamma i}(s, q, s_2), b^{1/s'}{}_{g\gamma'j}(s', q', s_2')^{\dagger}\} = \qquad (4.23)$$
$$= \Sigma_{s_1} \Sigma_{s_{1'}} (\int d^{r-1}p N'(p,q))^{\frac{1}{2}} (\int d^{r-1}p'N'(p',q))^{\frac{1}{2}} \{b^{1/s}{}_{\gamma i}(s,p,q,s_1,s_2), b^{1/s'}{}_{\gamma'j}(s', p', q', s_1', s_2')^{\dagger}\}$$
$$= (\int d^{r-1}pN'(p,q))^{\frac{1}{2}} (\int d^{r-1}p'N'(p',q))^{\frac{1}{2}} (1 - \delta_{ij}) \delta_{s,s'} \delta_{s_2 s_2'} \delta_{\gamma,\gamma'} \delta^{r-1}(\mathbf{p} - \mathbf{p}') \delta^{r-1}(\mathbf{q} - \mathbf{q}') U_g(\epsilon(s,q))$$
$$= \int d^{r-1}p N'(p,q) (1 - \delta_{ij}) \delta_{s,s'} \delta_{s_2,s_2'} \delta_{\gamma,\gamma'} \delta^{r-1}(\mathbf{p} - \mathbf{p}') \delta^{r-1}(\mathbf{q} - \mathbf{q}') U_g(\epsilon(s,q))$$
$$= \int d^{r-1}p N'(p',q) \delta^{r-1}(\mathbf{p} - \mathbf{p}') (1 - \delta_{ij}) \delta_{s,s'} \delta_{s_2,s_2'} \delta_{\gamma,\gamma'} \delta^{r-1}(\mathbf{q} - \mathbf{q}') U_g(\epsilon(s,q))$$
$$= (1 - \delta_{ij}) \delta_{s,s'} \delta_{s_2,s_2'} \delta_{\gamma,\gamma'} \delta^{r-1}(\mathbf{q} - \mathbf{q}') U_g(\epsilon(s,q))$$

and

$$\{d^{1/s}{}_{g\gamma i}(s, q, s_2), d^{1/s'}{}_{g\gamma'j}(s', q', s_2')^{\dagger}\} = (1 - \delta_{ij}) \delta_{s,s'} \delta_{s_2,s_2'} \delta_{\gamma,\gamma'} \delta^{r-1}(\mathbf{q} - \mathbf{q}') U_g(\epsilon(s,q))$$
$$(4.24)$$

where eq. 4.19 specifies $U_g(\epsilon(s,q))$ with $m = M$.

Eqs. 4.22 and 4.23 have fractional integrations that utilize

$$[(\int dp^{r-1} N'(p,q))^{\frac{1}{2}}]^2 = \int dp^{r-1} N'(p,q) \qquad (4.25)$$

as in Riemann-Liouville integrals. Note γ and γ' are internal symmetry indices.

Thus particles and gambols have the same internal symmetries and spin. They differ in mass and momentum. The corresponding d type operators have similar anti-commutation relations. The other anti-commutators are zero:

$$\{b^{1/s}{}_{g\gamma i}(s, q, s_2)^{\dagger}, b^{1/s'}{}_{g\gamma'j}(s', q', s_2')^{\dagger}\} = 0 \qquad (4.26)$$
$$\{d^{1/s}{}_{g\gamma i}(s, q, s_2)^{\dagger}, d^{1/s'}{}_{g\gamma'j}(s', q', s_2')^{\dagger}\} = 0$$
$$\{b^{1/s}{}_{g\gamma i}(s, q, s_2), b^{1/s'}{}_{g\gamma'j}(s', q', s_2')\} = 0$$
$$\{d^{1/s}{}_{g\gamma i}(s, q, s_2), d^{1/s'}{}_{g\gamma'j}(s', q', s_2')\} = 0$$

and the anti-commutators of b and d type operators are zero as well.

The gambol creation/annihilation operators above can be used to define gambol fermion quantum fields.

$$\psi_{g\gamma 1}(x) = \Sigma_p [b^{1/s}{}_{g\gamma 1}(s, p, s_1) f_p(x) + d^{1/s}{}_{g\gamma 1}(s, p, s_1)^{\dagger} f_p^*(x)] \qquad (4.27)$$
$$\psi_{g\gamma 2}(x) = \Sigma_p [b^{1/s}{}_{g\gamma 2}(s, p, s_1) f_p(x) + d^{1/s}{}_{g\gamma 2}(s, p, s_1)^{\dagger} f_p^*(x)]$$

4.3 External Particle Creation/Annihilation Operators

These operators may be defined using the composite creation/annihilation operators in eq. 4.14 for particles of momentum p. We replace[37] the sum over s[38] with a sum over energies ε so as to take advantage of the Planckian normalization sum:[39]

$$b_{\gamma i}(p, s_2) = \int_0^\infty d\varepsilon \; \Sigma_{s_1} \; (\int d^{r-1}q \; N'(p,q))^{\frac{1}{2}} \; b^{1/s}{}_{\gamma i}(s, p, q, s_1, s_2) \qquad (4.28)$$

$$b_{\gamma i}(p, s_2)^\dagger = \int_0^\infty d\varepsilon \; \Sigma_{s_1} \; (\int d^{r-1}q \; N'(p,q))^{\frac{1}{2}} \; b^{1/s}{}_{\gamma i}(s, p, q, s_1, s_2)^\dagger$$

The external particle anti-commutation relations are[40]

$$\{b_{\gamma i}(p, s_2), b_{\gamma' j}(p', s_2')^\dagger\} =$$
$$= \int d\varepsilon \int d\varepsilon' \; \Sigma_{s_1} \Sigma_{s_1'} (\int d^{r-1}q N'(p,q))^{\frac{1}{2}} (\int d^{r-1}q' N'(p,q))^{\frac{1}{2}} \{b^{1/s}{}_{\gamma i}(s,p,q,s_1,s_2), b^{1/s'}{}_{\gamma' j}{}^\dagger(s',p',q',s_1',s_2')\}$$
$$= \int d\varepsilon \int d\varepsilon' \Sigma_{s_1} \Sigma_{s_1'} (\int d^{r-1}q N'(p,q))^{\frac{1}{2}} (\int d^{r-1}q' N'(p,q))^{\frac{1}{2}} (1 - \delta_{ij}) \, \delta_{s_1,s_1'} \, \delta_{s_2,s_2'} \delta_{ss'} \delta_{\gamma\gamma'} \delta^{r-1}(\mathbf{p} - \mathbf{p'}) \cdot$$
$$\cdot \delta^{r-1}(\mathbf{q} - \mathbf{q'}) U_g(\varepsilon(s,q))$$
$$= \int d^{r-1}q \; N'(p,q) \, (1 - \delta_{ij}) \, \delta_{s_2,s_2'} \delta_{\gamma\gamma'} \delta^{r-1}(\mathbf{p} - \mathbf{p'}) \delta^{r-1}(\mathbf{q} - \mathbf{q'}) \int d\varepsilon \; U_g(\varepsilon(s,q))$$
$$= \int d^{r-1}q \; N'(p,q) \, \delta^{r-1}(\mathbf{q} - \mathbf{q'}) \, (1 - \delta_{ij}) \, \delta_{s_2,s_2'} \delta_{\gamma\gamma'} \delta^{r-1}(\mathbf{p} - \mathbf{p'})$$
$$= (1 - \delta_{ij}) \, \delta_{s_2,s_2'} \delta_{\gamma\gamma'} \delta^{r-1}(\mathbf{p} - \mathbf{p'}) \qquad (4.29)$$

using eqs. 4.19 and 4.20 with $\varepsilon(s)$, and $\varepsilon(s')$. We make the s (and ε) sums into integrations for analytic convenience just as the discrete sum over hν in the black body Planckian distribution derivation is similarly made continuous.

The other anti-commutation relations are:

$$\{d_{\gamma i}(p, s), d_{\gamma' j}(p', s')^\dagger\} = (1 - \delta_{ij}) \, \delta_{s,s'} \delta_{\gamma\gamma'} \delta^{r-1}(\mathbf{p} - \mathbf{p'}) \qquad (4.30)$$
$$\{b_{\gamma i}(p, s), b_{\gamma' j}(p', s')\} = 0$$
$$\{d_{\gamma i}(p, s), d_{\gamma' j}(p', s')\} = 0$$
$$\{b_{\gamma i}(p, s)^\dagger, b_{\gamma' j}(p', s')^\dagger\} = 0$$
$$\{d_{\gamma i}(p, s)^\dagger, d_{\gamma' j}(p', s')^\dagger\} = 0$$

and the anti-commutators of b and d type operators are zero.

The external particle[41] creation/annihilation operators above are those that appear in fermion quantum fields. Thus we may define an external *particle* quantum field with

---

[37] We use s and ε interchangeably.

[38] The sum over s begins as a discrete sum over powers of 2 according to Limos. We replace it with a continuous value for s, which makes the transition to an integral over ε possible.

[39] Using a sum over ε yields eq. 4.13.

[40] The s, s' and $\delta_{ss'}$ terms are equivalent to ε energy terms due to the 1:1 relation of s and ε.

[41] "External particle" means the particle as it appears as a particle without an interior part. The interior is masked by the external creation operator in eq. 4.28.

$$\psi_{\gamma 1}(x) = \Sigma_p \left[b_{\gamma 1}(p, s)\, f_p(x) + d_{\gamma 1}(p, s)^\dagger f_p^*(x) \right] \qquad (4.31)$$
$$\psi_{\gamma 2}(x) = \Sigma_p \left[b_{\gamma 2}(p, s) f_p(x) + d_{\gamma 2}(p, s)^\dagger f_p^*(x) \right]$$

with internal symmetry index γ.

5. Duplex Fermion Interior Tachyon PseudoFermion Quantum Fields

This chapter develops a two coordinate system, *duplex fermion,* formulation of fermion quantum field theory with a bradyon external external coordinate system and a tachyon internal coordinate system by using a new variant of *PseudoFermion*[42] quantum fields that we call Bradyon-Tachyon (BT) PseudoFermion quantum fields. The goal is to develop quantum field theories for both external and internal features of elementary particles. This chapter is based on Blaha (2018e) and (2022c) and section 4.4 of chapter 3.

It is often mentioned that a tachyon particle would rapidly decay – an instability giving tachyon condensation. The possibility that tachyons could lead to causality violation is also posed as a negative. The approach suggested in this book unites a bradyon part and a tachyon part in a fermion. As a result the tendency for tachyon condensation disappears. One has stable particles having an exterior bradyon part and tachyon interior part traveling at sub-light velocities.

We will also consider the possibility of a particle with a tachyon external part and bradyon internal part. The particle may exceed the speed of light. But it is stabilized by the interior bradyon part. The result is stable, superluminal particles. If such particles could be created in bulk they may support a starship engine for rapid travel to the stars.

We now turn to the consideration of tachyon-bradyon Physics in this and the following chapters.

5.1 BT PseudoFermions

We define[43] a BT PseudoFermion, with the form

$$\psi_{iBT\alpha\beta}(y, z) \tag{5.1}$$

where y and z are independent bradyon coordinates in r space-time dimensions, where i =1, 2 labels the PseudoQuantum fields, and α and β are spinor indices.[44]

---

[42] Chapter 6 of Blaha (2022c) discusses wave functions with multiple coordinates. See also (2023c).

[43] We follow the conventions of Bjorken (1965) with the $g^{\mu\nu}$ metric (1, -1, -1, -1).

[44] Appendix C of Blaha (2016f) presents a first quantized PseudoQuantum theory CQ Mechanics that embodies both classical and quantum theory. **This theory is the non-relativistic quantum mechanics limit of the PseudoFermion theory developed here.** CQ Mechanics has two sets of coordinates that combine to create a generalization of conventional Quantum Mechanics. It has applications in a generalized Feynman path integral formalism, a generalized Schrödinger equation, a generalized Boltzmann equation, the Fokker-Planck equation, a generalized approach to quantum and classical chaos, and to quantum entanglement as well as semi-quantum entanglement. Our "Pseudo" formalisms apply to both Quantum Field Theory and Quantum Mechanics. In these applications there is a clear almost continuous transition between the quantum to the classical sectors.

The BT Quantum field has a tachyon part for the z coordinates. The effective single tachyon free field equation for the z coordinates has the form:

$$(\gamma^\mu \partial/\partial(iz)^\mu - m)\psi_T(z) = (\gamma^\mu \partial/\partial z^\mu - m)\psi_T(z) = 0 \qquad (3.33)$$

where z is viewed as a set of imaginary coordinates.

The "missing" factor of i in the first term of eq. 3.33 requires the Lagrangian to be different from the conventional Dirac Lagrangian in order for the Lagrangian to be real. The simplest, physically acceptable, free spin ½ tachyon Lagrangian density for a one coordinates case is:

$$\mathcal{L}_T = \psi_T{}^S(\gamma^\mu \partial/\partial z^\mu - m)\psi_T(z) \qquad (3.34)$$

where

$$\psi_T{}^S = \psi_T{}^\dagger \, i\gamma^0\gamma^5 \qquad (3.35)$$

The corresponding action is

$$I = \int d^4 z \, \mathcal{L}_T \qquad (3.36)$$

Appendix 3-B of Blaha (2007b) proves I is real.

We return now to the free PseudoFermion PseudoQuantum field with two sets of coordinates.

We begin by defining a free PseudoFermion PseudoQuantum Lagrangian with two related quantum fields ψ_{1BT} and ψ_{2BT} that are functions of the two sets of coordinates in r space-time dimensions, y and z with the same metric tensor: $g_{\mu\nu y} = \mathrm{diag}(1, -1, -1, -1) = g_{\mu\nu z}$. The y coordinates are "bradyon" coordinates. The z coordinates are "tachyon" coordinates. Thus we implement a form of complex space-time and define Dirac matrices correspondingly: $\gamma_{y\alpha\kappa}{}^\mu$ and $\gamma_{z\beta\lambda}{}^\nu$

$$\mathcal{L} = \overline{\psi}_{2BT}{}^{\alpha\beta}[iM^{-1}\gamma_{y\alpha\kappa}{}^\mu \partial/\partial y^\mu \gamma_{z\beta\lambda}{}^\nu \partial/\partial z^\nu - M]\psi_{1BT}{}^{\kappa\lambda} +$$
$$+ \overline{\psi}_{1BT}{}^{\alpha\beta}[iM^{-1}\gamma_{y\alpha\kappa}{}^\mu \partial/\partial y^\mu \, \gamma_{z\beta\lambda}{}^\nu \partial/\partial z^\nu - M]\psi_{2BT}{}^{\kappa\lambda} \qquad (5.2)$$

where $\gamma_y{}^\mu$ and $\gamma_z{}^\mu$ are Dirac matrices for the y and z coordinates respectively, and y and z are coordinates in r dimension space-times, M is the mass, and

$$\overline{\psi}_{kBT\alpha\beta} = \psi_{kBT}{}^{\kappa\lambda\dagger}\gamma_y{}^0{}_{\kappa\alpha} \, i(\gamma_z{}^0\gamma_z{}^5)_{\lambda\beta} \qquad (5.2')$$

for i, k = 1, 2 where the subscripts y and y indicate Dirac spinors associated with the y and z coordinates.

The form of the complex conjugate, eq. 5.2', implies that the alternate possible expression (a summation) for $\mathcal{L}$, namely $i\gamma_{y\alpha\kappa}{}^\mu \partial/\partial y^\mu + \gamma_{z\beta\lambda}{}^\nu \partial/\partial z^\nu$ is not appropriate.

The equations of motion are

$$[iM^{-1}\gamma_y{}^{\mu}\partial/\partial y^{\mu}\ \gamma_z{}^{\nu}\partial/\partial z^{\nu} - M]\psi_{1BT} = 0 \qquad (5.2a)$$
$$[iM^{-1}\gamma_y{}^{\mu}\cdot\partial/\partial y^{\mu}\ \gamma_z{}^{\nu}\partial/\partial z^{\nu} - M]\psi_{2BT} = 0 \qquad (5.2b)$$

We define subsidiary equations of motion

$$[i\gamma_y{}^{\mu}\partial/\partial y^{\mu} - M]\psi_{jBT} = 0 \qquad (5.2c)$$
$$[\gamma_z{}^{\nu}\partial/\partial z^{\nu} - M]\psi_{jBT} = 0 \qquad (5.2d)$$

for j = 1, 2. Eqs. 5.2c and 5.2d imply eqs. 5.2a and 5.2b. Note eq. 5.2d is tachyonic by design.

One conjugate momentum is (with two spinor indices α, β)

$$\pi_{y1BT\alpha\beta} = \partial \mathscr{L}/\partial(\partial\psi_{1BT\alpha\beta}/\partial y^{0}) = M^{-1}(\partial/\partial z^{\nu}\ \psi_{2BT}{}^{\dagger\alpha\lambda}(\gamma_z{}^{0}\gamma_z{}^{5}\gamma_z{}^{\nu})_{\lambda\beta} = (\psi_{2BT}{}^{\dagger\alpha\lambda}(\gamma_z{}^{0}\gamma_z{}^{5})_{\lambda\beta} \quad (5.3)$$

after partial integrations (with surface terms set to zero) of

$$L = \int d^r y \int d^r z\ \mathscr{L}$$

using the subsidiary tachyonic equation of motion:

$$\gamma_z{}^{\nu}\partial/\partial z^{\nu}\ \psi_{2BT}{}^{\dagger} = M\psi_{2BT}{}^{\dagger} \qquad (5.4)$$

Similarly the other conjugate momenta are

$$\pi_{z1BT\alpha\beta} = \partial \mathscr{L}/\partial(\partial\psi_{1BT\alpha\beta}/\partial z^{0}) = -i\psi_{2BT}{}^{\dagger\kappa\lambda}\gamma_y{}^{0}{}_{\kappa\alpha}\ \gamma_z{}^{5}{}_{\lambda\beta} \qquad (5.5)$$
$$\pi_{y2BT\alpha\beta} = \partial \mathscr{L}/\partial(\partial\psi_{2BT\alpha\beta}/\partial y^{0}) = \psi_{1BT}{}^{\dagger\alpha\lambda}(\gamma_z{}^{0}\gamma_z{}^{5})_{\lambda\beta} \qquad (5.6)$$
$$\pi_{z2BT\alpha\beta} = \partial \mathscr{L}/\partial(\partial\psi_{2BT\alpha\beta}/\partial z^{0}) = -i\psi_{1BT}{}^{\dagger\kappa\lambda}\gamma_y{}^{0}{}_{\kappa\alpha}\ \gamma_z{}^{5}{}_{\lambda\beta} \qquad (5.7)$$

using the subsidiary equations of motion:

$$\gamma_z{}^{\nu}\partial/\partial z^{\nu}\ \psi_{1BT}{}^{\dagger} = M\psi_{1BT}{}^{\dagger} \qquad (5.8)$$
$$i\gamma_y{}^{\nu}\partial/\partial y^{\nu}\ \psi_{1BT}{}^{\dagger} = M\psi_{1BT}{}^{\dagger} \qquad (5.9)$$
$$i\gamma_y{}^{\nu}\partial/\partial y^{\nu}\ \psi_{2BT}{}^{\dagger} = M\psi_{2BT}{}^{\dagger} \qquad (5.10)$$

We define new momenta to have a y − z symmetry:

$$\pi_{1BT\alpha\beta} = (\pi_{y1BT}\gamma_z{}^{5}\gamma_z{}^{0})_{\alpha\beta} = (\pi_{z1BT}\gamma_z{}^{5}\gamma_y{}^{0})_{\alpha\beta} = (\psi_{2BT}{}^{\dagger})_{\alpha\beta} \qquad (5.11)$$
$$\pi_{2BT\alpha\beta} = (\pi_{y2BT}\gamma_z{}^{0}\gamma_z{}^{5})_{\alpha\beta} = (\pi_{z2BT}\gamma_z{}^{5}\gamma_y{}^{0})_{\alpha\beta} = (\psi_{1BT}{}^{\dagger})_{\alpha\beta} \qquad (5.12)$$

where the terms symbolically represent those in eqs. 5.3, 5.5, 5.6 and 5.7.
The form of the conjugate momenta gives the non-zero equal time anticommutators: [45]

$$\{\pi_{jBT\alpha\beta}(\mathbf{y}, y^0, \mathbf{z}, z^0), \psi_{iBT\kappa\lambda}(\mathbf{y'}, y^0, \mathbf{z'}, z^{0'})\} = \{\psi_{jBT\alpha\beta}^{\dagger}(\mathbf{y}, y^0, \mathbf{z}, z^0), \psi_{iBT\kappa\lambda}(\mathbf{y'}, y^0, \mathbf{z'}, z^{0'})\}$$

$$= (1 - \delta_{ij})\delta_{\alpha\kappa}\delta_{\beta\lambda}\,\delta^{r-1}(\mathbf{y} - \mathbf{y'})\delta^{r'-1}(z - \mathbf{z'}) \qquad (5.13)$$

$$\{\pi_{jBT\alpha\beta}(\mathbf{y}, y^0, \mathbf{z}, z^0), \psi_{iBT\kappa\lambda}(\mathbf{y'}, y^{0'}, \mathbf{z'}, z^0)\} = \{\psi_{jBT\alpha\beta}^{\dagger}(\mathbf{y}, y^0, \mathbf{z}, z^0), \psi_{iBT\kappa\lambda}(\mathbf{y'}, y^{0'}, \mathbf{z'}, z^0)\}$$

$$= (1 - \delta_{ij})\delta_{\alpha\kappa\,\beta}\delta_{\beta\lambda}\,\delta^{r-1}(\mathbf{y} - \mathbf{y'})\delta^{r'-1}(z - \mathbf{z'})$$

for i, j = 1, 2.

5.2 PseudoFermion Quantum Fields with a Tachyon Interior

A free PseudoFermion wave function with a tachyon interior has the form:

$$\psi_{iBT\alpha\beta}(y, z) = \sum_{s_1}\sum_{s_2} \int dp^{r-1}\int dq^{r-1}\, N(p, q)\, [b_{iBT}(p, q, s_1, s_2)u_\alpha(p, s_1)u_{\beta BT}(q, s_2)\exp(-ip\cdot y - iq\cdot z) +$$

$$+ d_{iBT}^{\dagger}(p, q, s1, s2)v_\alpha(p, s1)v_{\beta BT}(q, s2)\exp(ip\cdot y + iq\cdot z)] \qquad (5.14)$$

plus Hermitean conjugates for i = 1, 2 where $N(p, q)$ is a normalization factor. Note both parts (y and z) of ψ_i are on the mass shell: $p^2 = M^2$ and $q^2 = -M^2$.

At this point we might attempt to complete the canonical quantization procedure in the conventional manner by Fourier expanding the quantum field and specifying anti-commutation relations for the Fourier component amplitudes. However the incompleteness of the set of plane waves, which are limited by the restriction $|q| \geq M$, causes the anti-commutator of the fields not to yield a $\delta^3(x - x')$. Thus the conventional approach fails to yield the required anti-commutation relations.

Other approaches: 1) decompose the tachyon field into left-handed and right-handed parts and then second quantize each part; and 2) second quantize in light-front coordinates ($x^{\pm} = (x^0 \pm x^3)/\sqrt{2}$). These approaches also both fail.[46]

The only approach that does succeed[47] is to decompose the tachyon part of the field into left-handed and right-handed parts and then second quantize the z coordinate part in light-front coordinates. We follow this procedure in the following subsections.

5.2.1 Separation into Left-Handed and Right-Handed Fields

We will use a transformed set of Dirac matrices to develop our left-handed and right-handed tachyon formulations:

---

[45] See S. Blaha, Il Nuovo Cimento **49A**, 35 (1979) for one coordinate system, PseudoQuantum fermions.
[46] See the first edition Blaha (2006) where these possibilities were considered and found to fail.
[47] Blaha (2006) discusses this case in detail.

$$\gamma^0 = \begin{bmatrix} 0 & -I \\ -I & 0 \end{bmatrix} \qquad \gamma^i = \begin{bmatrix} 0 & \sigma_i \\ -\sigma_i & 0 \end{bmatrix} \qquad \gamma^5 = \begin{bmatrix} I & 0 \\ 0 & -I \end{bmatrix}$$

$$(5.15)$$

which are obtained from the usual Dirac matrices by applying the unitary transformation $U_z = 2^{1/2}(I + \gamma_z^5\gamma_z^0)$. *I is the 4×4 identity matrix.* The γ_z^5 chirality operator's eigenvalues define handedness: +1 corresponds to right-handed; and −1 corresponds to left-handed:

$$\gamma_z^5\psi_{BTL} = -\,\psi_{BTL} \qquad\qquad \gamma_z^5\psi_{BTR} = \psi_{BTR} \qquad (5.16)$$

Consequently, we can define left-handed and right-handed tachyon fields with the projection operators in eqs. 3.49 and 3.50.

We can calculate the commutation relations of the resulting left-handed and right-handed fields by pre-multiplying and post-multiplying by $\frac{1}{2}(1 - \gamma_z^5)$ and $\frac{1}{2}(1 + \gamma_z^5)$. The results are:

$$\{\pi_{jBTL\alpha\beta}(\mathbf{y}, y^0,\mathbf{z},z^0), \ \psi_{iBTL\kappa\lambda}(\mathbf{y'}, y^0, \mathbf{z'}, z^{0\prime})\} = \{ \ \psi_{jBTL\alpha\beta}^\dagger(\mathbf{y}, y^0,\mathbf{z},z^0), \ \psi_{iBTL\kappa\lambda}(\mathbf{y'}, y^0, \mathbf{z'}, z^{0\prime})\}$$
$$= 1/4 \ (1 - \delta_{ij})(1 - \gamma^5)_{\alpha\kappa}(1 - \gamma^5)_{\beta\lambda}\,\delta^{r-1}(\mathbf{y} - \mathbf{y'})\delta^{r-1}(\mathbf{z} - \mathbf{z'}) \quad (5.17)$$

$$\{\pi_{jBTR\alpha\beta}(\mathbf{y}, y^0,\mathbf{z},z^0), \ \psi_{iBTR\kappa\lambda}(\mathbf{y'}, y^0,\mathbf{z'},z^{0\prime})\} = \{ \ \psi_{jBTR\alpha\beta}^\dagger(\mathbf{y}, y^0,\mathbf{z},z^0), \ \psi_{iBTR\kappa\lambda}(\mathbf{y'}, y^0,\mathbf{z'},z^{0\prime})\}$$

$$= -1/4 \ (1 - \delta_{ij})(1 + \gamma^5)_{\alpha\kappa}(1 + \gamma^5)_{\beta\lambda}\,\delta^{r-1}(\mathbf{y} - \mathbf{y'})\delta^{r-1}(\mathbf{z} - \mathbf{z'}) \quad (5.18)$$

$$\{\pi_{jBTL\alpha\beta}(\mathbf{y}, y^0,\mathbf{z},z^0), \ \psi_{iBTR\kappa\lambda}(\mathbf{y'}, y^0,\mathbf{z'},z^{0\prime})\} = \{\psi_{jBTL\alpha\beta}^\dagger(\mathbf{y}, y^0,\mathbf{z},z^0), \ \psi_{iBTR\kappa\lambda}(\mathbf{y'}, y^0,\mathbf{z'},z^{0\prime})\} = 0$$
$$(5.19)$$

$$\{\pi_{jBTR\alpha\beta}(\mathbf{y}, y^0,\mathbf{z},z^0), \ \psi_{iBTL\kappa\lambda}(\mathbf{y'}, y^0,\mathbf{z'},z^{0\prime})\} = \{\psi_{jBTR\alpha\beta}^\dagger(\mathbf{y}, y^0,\mathbf{z},z^0), \ \psi_{iBTL\kappa\lambda}(\mathbf{y'}, y^0,\mathbf{z'},z^{0\prime})\} = 0$$
$$(5.20)$$

and similar expressions for the equal z^0 case.

We decompose the Lagrangian into parts using eq. 3.50. The Lagrangian density above decomposes into left-handed and right-handed parts. Eq. 5.2 becomes

$$\begin{aligned}
\mathcal{L} = &-\psi_{2BTL}^{\dagger\alpha\beta}M^{-1}(\gamma_y^0\gamma_y^\mu)_{\alpha\kappa}\partial/\partial y^\mu(\gamma_z^0\gamma_z^\nu)_{\beta\lambda}\partial/\partial z^\nu\psi_{1BTL}^{\kappa\lambda} - iM\psi_{2BTR}^{\dagger\alpha\beta}\gamma_y^0{}_{\alpha\kappa}\gamma_z^0{}_{\beta\lambda}\psi_{1BTL}^{\kappa\lambda} + \\
&+ \psi_{2BTR}^{\dagger\alpha\beta}M^{-1}(\gamma_y^0\gamma_y^\mu)_{\alpha\kappa}\partial/\partial y^\mu(\gamma_z^0\gamma_z^\nu)_{\beta\lambda}\partial/\partial z^\nu\psi_{1BTR}^{\kappa\lambda} + iM\psi_{2BTL}^{\dagger\alpha\beta}\gamma_y^0{}_{\alpha\kappa}\gamma_z^0{}_{\beta\lambda}\psi_{1BTR}^{\kappa\lambda} - \\
&- \psi_{1BTL}^{\dagger\alpha\beta}M^{-1}(\gamma_y^0\gamma_y^\mu)_{\alpha\kappa}\partial/\partial y^\mu(\gamma_z^0\gamma_z^\nu)_{\beta\lambda}\partial/\partial z^\nu\psi_{2BTL}^{\kappa\lambda} - iM\psi_{1BTR}^{\dagger\alpha\beta}\gamma_y^0{}_{\alpha\kappa}\gamma_z^0{}_{\beta\lambda}\psi_{2BTL}^{\kappa\lambda} + \\
&+ \psi_{1BTR}^{\dagger\alpha\beta}M^{-1}(\gamma_y^0\gamma_y^\mu)_{\alpha\kappa}\partial/\partial y^\mu(\gamma_z^0\gamma_z^\nu)_{\beta\lambda}\partial/\partial z^\nu\psi_{2BTR}^{\kappa\lambda} + iM\psi_{1BTL}^{\dagger\alpha\beta}\gamma_y^0{}_{\alpha\kappa}\gamma_z^0{}_{\beta\lambda}\psi_{2BTR}^{\kappa\lambda}
\end{aligned}$$

$$(5.21)$$

5.3 Further Separation into + and – Light-Front Fields

There have been many studies of light-front (infinite momentum frame) physics in the past forty years. Light-front coordinates *cannot* be obtained by a Lorentz transformation, or by a superluminal transformation, from a standard set of coordinate system variables even in a limiting sense. Instead they are a defined set of variables that have been used to develop quantum field theories that have been shown to be equivalent to quantum field theories based on conventional coordinates. In particular, light-front quantum field theories have been shown to yield fully Lorentz covariant S matrix elements that are the same as S matrix elements calculated in the conventional way.

Light-front variables for z coordinates can be defined by:

$$z^{\pm} = (z^0 \pm z^3)/\sqrt{2} \tag{5.22}$$
$$\partial/\partial z^{\pm} \equiv \partial^{\mp} \equiv (\partial/\partial z^0 \pm \partial/\partial z^3)/\sqrt{2}$$

with the "transverse" coordinate variables, z^1 and z^2, unchanged.

The inner product of two 4-vectors has the form

$$x{\cdot}z = x^+z^- + z^+x^- - x^1z^1 - x^2z^2 \tag{5.23}$$

and the light-front definition of Dirac matrices is:

$$\gamma^{\pm} = (\gamma^0 \pm \gamma^3)/\sqrt{2} \tag{5.24}$$

with transverse matrices γ^1 and γ^2 defined as usual. Note the useful identity:

$$\gamma^{\pm 2} = 0$$

We define "+" and "–" tachyon fields with the projection operators:

$$R^{\pm} = \tfrac{1}{2}(I \pm \gamma^0\gamma^3) \tag{5.25}$$

and C where C is defined in eqs. 3.49 and 3.50. They are:

Left-handed, ± light-front fields: $\psi_{BTL}{}^{\pm} = R^{\pm}C^{-}\psi_{BT}$

Right-handed, ± light-front fields: $\psi_{BTR}{}^{\pm} = R^{\pm}C^{+}\psi$
$$\tag{5.26}$$

We define

$$\Psi_{iBTL\alpha}{}^{\lambda} = i(\gamma_y{}^0\gamma_y)_{\alpha\kappa}{}^{\mu}\partial/\partial y^{\mu}\psi_{iBTL}{}^{\kappa\lambda} = M\,\gamma_y{}^0{}_{\alpha\kappa}\psi_{iBTL}{}^{\kappa\lambda} \tag{5.27}$$
$$\Psi_{iBTR\alpha}{}^{\lambda} = i(\gamma_y{}^0\gamma_y)_{\alpha\kappa}{}^{\mu}\partial/\partial y^{\mu}\psi_{iBTR}{}^{\kappa\lambda} = M\,\gamma_y{}^0{}_{\alpha\kappa}\psi_{iBTR}{}^{\kappa\lambda}$$

for convenience with PseudoQuantum $i = 1, 2$ using the bradyon subsidiary conditions

$$i\gamma_y{}^\nu \partial/\partial y^\nu \, \psi_{iBTL} = M\psi_{iBTL} \qquad (5.28)$$
$$i\gamma_y{}^\nu \partial/\partial y^\nu \, \psi_{iBTR} = M\psi_{iBTR}$$

Then eq. 5.21 becomes

$$
\begin{aligned}
\mathcal{L} = {}&i\psi_{2BTL}{}^{\dagger\alpha\beta}M^{-1}(\gamma_z{}^0\gamma_z{}^\nu)_{\beta\lambda}\partial/\partial z^\nu\Psi_{1BTL\alpha}{}^\lambda - iM\psi_{2BTR}{}^{\dagger\alpha\beta}\gamma_y{}^0{}_{\alpha\kappa}\gamma_z{}^0{}_{\beta\lambda}\Psi_{1BTL}{}^{\kappa\lambda} - \qquad (5.29)\\
&- i\psi_{2BTR}{}^{\dagger\alpha\beta}M^{-1}(\gamma_z{}^0\gamma_z{}^\nu)_{\beta\lambda}\partial/\partial z^\nu\Psi_{1BTR\alpha}{}^\lambda + iM\psi_{2BTL}{}^{\dagger\alpha\beta}\gamma_y{}^0{}_{\alpha\kappa}\gamma_z{}^0{}_{\beta\lambda}\Psi_{1BTR}{}^{\kappa\lambda} + \\
&+ i\psi_{1BTL}{}^{\dagger\alpha\beta}M^{-1}(\gamma_z{}^0\gamma_z{}^\nu)_{\beta\lambda}\partial/\partial z^\nu\Psi_{2BTL\alpha}{}^\lambda - iM\psi_{1BTR}{}^{\dagger\alpha\beta}\gamma_y{}^0{}_{\alpha\kappa}\gamma_z{}^0{}_{\beta\lambda}\Psi_{2BTL}{}^{\kappa\lambda} - \\
&- i\psi_{1BTR}{}^{\dagger\alpha\beta}M^{-1}(\gamma_z{}^0\gamma_z{}^\nu)_{\beta\lambda}\partial/\partial z^\nu\Psi_{2BTR\alpha}{}^\lambda + iM\psi_{1BTL}{}^{\dagger\alpha\beta}\gamma_y{}^0{}_{\alpha\kappa}\gamma_z{}^0{}_{\beta\lambda}\Psi_{2BTR}{}^{\kappa\lambda}
\end{aligned}
$$

Or using eq. 5.27

$$
\begin{aligned}
\mathcal{L} = {}&i\psi_{2BTL}{}^{\dagger\alpha\beta}\gamma_y{}^0{}_{\alpha\kappa}(\gamma_z{}^0\gamma_z{}^\nu)_{\beta\lambda}\partial/\partial z^\nu\psi_{1BTL}{}^{\kappa\lambda} - iM\psi_{2BTR}{}^{\dagger\alpha\beta}\gamma_y{}^0{}_{\alpha\kappa}\gamma_z{}^0{}_{\beta\lambda}\psi_{1BTL}{}^{\kappa\lambda} - \qquad (5.30)\\
&- i\psi_{2BTR}{}^{\dagger\alpha\beta}\gamma_y{}^0{}_{\alpha\kappa}(\gamma_z{}^0\gamma_z{}^\nu)_{\beta\lambda}\partial/\partial z^\nu\psi_{1BTR}{}^{\kappa\lambda} + iM\psi_{2BTL}{}^{\dagger\alpha\beta}\gamma_y{}^0{}_{\alpha\kappa}\gamma_z{}^0{}_{\beta\lambda}\psi_{1BTR}{}^{\kappa\lambda} + \\
&+ i\psi_{1BTL}{}^{\dagger\alpha\beta}\gamma_y{}^0{}_{\alpha\kappa}(\gamma_z{}^0\gamma_z{}^\nu)_{\beta\lambda}\partial/\partial z^\nu\psi_{2BTL}{}^{\kappa\lambda} - iM\psi_{1BTR}{}^{\dagger\alpha\beta}\gamma_y{}^0{}_{\alpha\kappa}\gamma_z{}^0{}_{\beta\lambda}\psi_{2BTL}{}^{\kappa\lambda} - \\
&- i\psi_{1BTR}{}^{\dagger\alpha\beta}\gamma_y{}^0{}_{\alpha\kappa}(\gamma_z{}^0\gamma_z{}^\nu)_{\beta\lambda}\partial/\partial z^\nu\psi_{2BTR}{}^{\kappa\lambda} + iM\psi_{1BTL}{}^{\dagger\alpha\beta}\gamma_y{}^0{}_{\alpha\kappa}\gamma_z{}^0{}_{\beta\lambda}\psi_{2BTR}{}^{\kappa\lambda}
\end{aligned}
$$

We now surpress α, β, and λ indices with the y coordinates indices understood and then transform to light-front variables and fields as above we obtain the light-front free tachyon Lagrangian:

$$
\begin{aligned}
\mathcal{L} = {}&i\psi_{2BTL}{}^\dagger\gamma_y{}^0\gamma_z{}^0\gamma_z{}^\nu\partial/\partial z^\nu\Psi_{1BTL} - iM\psi_{2BTR}{}^\dagger\gamma_y{}^0\gamma_z{}^0\Psi_{1BTL} - \qquad (5.31)\\
&- i\psi_{2BTR}{}^\dagger\gamma_y{}^0\gamma_z{}^0\gamma_z{}^\nu\partial/\partial z^\nu\Psi_{1BTR} + iM\psi_{2BTL}{}^\dagger\gamma_y{}^0\gamma_z{}^0\Psi_{1BTR} + \\
&+ i\psi_{1BTL}{}^\dagger\gamma_y{}^0\gamma_z{}^0\gamma_z{}^\nu\partial/\partial z^\nu\Psi_{2BTL} - iM\psi_{1BTR}{}^\dagger\gamma_y{}^0\gamma_z{}^0\Psi_{2BTL} - \\
&- i\psi_{1BTR}{}^\dagger\gamma_y{}^0\gamma_z{}^0\gamma_z{}^\nu\partial/\partial z^\nu\Psi_{2BTR} + iM\psi_{1BTL}{}^\dagger\gamma_y{}^0\gamma_z{}^0\Psi_{2BTR}
\end{aligned}
$$

The form of eq. 3.60 results in

$$
\begin{aligned}
\mathcal{L} = {}&-2^{\frac12}\psi_{2BTL}{}^{++}\gamma_y{}^0\gamma_y{}^\mu\partial/\partial y^\mu\partial_z{}^-\Psi_{1BTL}{}^+ - 2^{\frac12}\psi_{2BTL}{}^{-\dagger}\gamma_y{}^0\gamma_y{}^\mu\partial/\partial y^\mu\partial_z{}^+\Psi_{1BTL}{}^- + \\
&+ \psi_{2BTL}{}^{+\dagger}\gamma_z{}^0\gamma_y{}^0\gamma_y{}^\mu\partial/\partial y^\mu\gamma_z{}^j\partial^j{}_z\Psi_{1BTL}{}^- + \psi_{2BTL}{}^{-\dagger}\gamma_z{}^0\gamma_y{}^0\gamma_y{}^\mu\partial/\partial y^\mu\gamma_z{}^j\partial_z{}^j\Psi_{1BTL}{}^+ + \\
&+ 2^{\frac12}\psi_{2BTR}{}^{+\dagger}\gamma_z{}^0\gamma_y{}^0\gamma_y{}^\mu\partial/\partial y^\mu\partial_z{}^-\Psi_{1BTR}{}^+ + 2^{\frac12}\psi_{2BTR}{}^{-\dagger}\gamma_z{}^0\gamma_y{}^0\gamma_y{}^\mu\partial/\partial y^\mu\partial_z{}^+\Psi_{1BTR}{}^- - \\
&- \psi_{2BTR}{}^{+\dagger}\gamma_z{}^0\gamma_y{}^0\gamma_y{}^\mu\partial/\partial y^\mu\gamma_z{}^j\partial_z{}^j\Psi_{1BTR}{}^- - \psi_{2BTR}{}^{-\dagger}\gamma_z{}^0\gamma_y{}^0\gamma_y{}^\mu\partial/\partial y^\mu\gamma_z{}^j\partial_z{}^j\Psi_{1BTR}{}^+ - \\
&- iM[\psi_{2BTR}{}^{+\dagger}\gamma_z{}^0\gamma_y{}^0\Psi_{1BT_L}{}^- - \psi_{2BTL}{}^{+\dagger}\gamma_z{}^0\gamma_y{}^0\Psi_{1BTR}{}^- + \psi_{2BTR}{}^{-\dagger}\gamma_z{}^0\gamma_y{}^0\Psi_{1BTL}{}^+ - \\
&- \psi_{2BTL}{}^{-\dagger}\gamma_z{}^0\gamma_y{}^0\Psi_{1BTR}{}^+] +
\end{aligned}
$$

$$+ (1 \leftrightarrow 2) \qquad (5.32)$$

     Chapter 3 contains more detail on the tachyon sector. Please read from eq. 3.62 to eq, 3.70.

5.4 Left-Handed Tachyons

We now turn to the tachyonic sector within the external bradyon sector of the PseudoFermion fields. We view a fundamental fermion as having an external part with y coordinates and an internal part with z coordinates that may be bradyonic or tachyonic.

If we introduce fractionation[48] s where the particle is fractionated to s gambols, and has internal symmetry index γ, the free "+" light-front, left-handed, PseudoFermion wave function has the form:

$$\psi_{iBT\gamma\alpha L}{}^{+}(y, z) = \Sigma\Sigma\Sigma \int dp^{r-1}\int d^{r-2}qdq^{+} N_L{}^{+}(p, q)\, \theta(q^{+})\, [b_{iBT\gamma L}{}^{+}(s, p, q, s_1, s_2) \cdot$$
$$\cdot u_\alpha(p,s_1)u_{BTL}{}^{+}(q,s_2)\exp(-ip\cdot y - iq\cdot z) +$$
$$+ d_{iBT\gamma L}{}^{++}(s, p, q, s_1, s_2)^{\dagger}v_\alpha(p,s_1)v_{BTL}{}^{+}(q,s_2)\exp(ip\cdot y + iq\cdot z)]$$
$$(5.33)$$

following eq. 3.71 plus Hermitean conjugates for $i = 1, 2$ with a spinor index α, where s_i is the spin and $N_L{}^{+}(p, q)$ is a normalization factor. Its Hermitean conjugate is

$$\psi_{iBT\gamma\alpha L}{}^{++}(y, z) = \Sigma\Sigma\Sigma \int dp^{r-1}\int d^{r-2}qdq^{+} N_L{}^{+}(p, q)\, \theta(q^{+})\cdot$$
$$\cdot [b_{iBT\gamma L}{}^{++}(s,p,q,s_1,s_2)u_\alpha{}^{\dagger}(p,s_1)u_{BTL}{}^{++}(q,s_2)\exp(ip\cdot y + iq\cdot z) +$$
$$+ d_{iBT\gamma L}{}^{+}(s, p, q, s_1, s_2)^{\dagger}v_\alpha{}^{\dagger}(p,s_1)v_{BTL}{}^{++}(q,s_2)\exp(-ip\cdot y - iq\cdot z)]$$
$$(5.34)$$

where † indicates Hermitean conjugate.

The anti-commutation relations of the Fourier coefficient operators are

$$\{b_{iBT\gamma L}{}^{+}(s, p, q, s_1, s_2), b_{jBT\gamma'L}{}^{++}(s', p', q', s_1', s_2')\} =$$
$$= (1 - \delta_{ij})\delta_{\gamma\gamma'}\,\delta_{s,s'}\delta_{s_1,s_1'}\delta_{s_2,s_2'}\delta^{r-1}(\mathbf{p - p'})\delta^{r-2}(\mathbf{q - q'})\delta(q^{+} - q'^{+}))\, U_g(\epsilon(s, p\bullet q/M)) \quad (5.35)$$

$$\{d_{iBT\gamma L}{}^{+}(s, p, q, s_1, s_2), d_{jBT\gamma'L}{}^{++}(s', p', q', s_1', s_2')\} =$$
$$= (1 - \delta_{ij})\delta_{\gamma\gamma'}\delta_{s,s'}\delta_{s_1,s_1'}\delta_{s_2,s_2'}\delta^{r-1}(\mathbf{p - p'})\delta^{r-2}(\mathbf{q - q'})\delta(q^{+} - q'^{+}))\, U_g(\epsilon(s, p\bullet q/M)) \quad (5.36)$$

where U_g is given by the gambol *Planckian* distribution *for particles*:[49]

$$U_g(\epsilon) = 15\, N\, (\pi kT)^{-4}\, \epsilon^{3}/(e^{\epsilon/kT} - 1) \quad (5.37)$$

where N_L is the normalization constant and

$$\epsilon(s.\, p\bullet q/M) = [(m_g s + M)/(s + 1)]p\bullet q/M \quad (5.38)$$

where s is the fractionation parameter, and m_g is the gambol mass.

---

[48] The fractionation into gambols may be easily omitted.
[49] Blaha (2024b).

The other Left anti-commutators are zero (as in eq. 3.74). The tachyonic spinors have the form of eq. 3.75 and Appendix 3-A.

5. 5 Right-Handed Tachyons

The case of the right-handed tachyon part of a particle is similar to the left-handed case with only two differences: a minus sign in the creation and annihilation operator anti-commutation relations, and the use of right-handed projection operators. The right-handed tachyon wave function light-front Fourier expansion is:

$$\psi_{iBT\gamma\alpha R}{}^+(y,z) = \Sigma\Sigma\Sigma_{s\ s_1\ s_2} \int dp^{r-1}\int d^{r-2}qdq^+ \ N_R{}^+(p,q)\ \theta(q^+)\ [b_{iBT\gamma R}{}^+(s,p,q,s_1,s_2)u_\alpha(p,s_1)u_{BTR}{}^+(q,s_2)\bullet$$

$$\bullet\exp(-ip\cdot y - iq\cdot z) + d_{iBT\gamma R}{}^{++}(s,\ p,\ q,\ s_1,\ s_2)^\dagger v_\alpha(p,s_1)v_{BTR}{}^+(q,s_2)\exp(ip\cdot y + iq\cdot z)]$$

$$(5.39)$$

following eq. 3.71 plus Hermitean conjugates for $i = 1, 2$ with a spinor index α, where s_i is the spin and $N_R{}^+(p, q)$ is a normalization factor. Its Hermitean conjugate is

$$\psi_{iBT\gamma\alpha R}{}^{++}(y,\ z) = \Sigma\Sigma\Sigma_{s\ s_1\ s_2} \int dp^{r-1}\int d^{r-2}qdq^+ \ N_R{}^+(p,\ q)\theta(q^+)\ [b_{iBT\gamma R}{}^{++}(s,p,q,s_1,s_2)u_\alpha{}^\dagger(p,s_1)\bullet$$

$$\bullet u_{BTR}{}^{++}(q,s_2)\exp(ip\cdot y + iq\cdot z) +$$

$$+ d_{iBT\gamma R}{}^+(s,\ p,\ q,\ s_1,\ s_2)^\dagger v_\alpha{}^\dagger(p,s_1)v_{BTR}{}^{++}(q,s_2)\exp(-ip\cdot y - iq\cdot z)]$$

$$(5.40)$$

where † indicates Hermitean conjugate.

We assume the anti-commutation relations of the Fourier coefficient operators for the gambol interior are[50]

$$\{b_{iBT\gamma_R}{}^+(\ s,\ p,\ q,\ s_1,\ s_2),\ b_{jBT\gamma'_R}{}^{++}(\ s',\ p',\ q',\ s_1',\ s_2')\} =$$
$$= -(1 - \delta_{ij})\delta_{\gamma\gamma'}\delta_{s,s'}\delta_{s_1,s_1'}\delta_{s_2,s_2'}\ \delta^{r-1}(\mathbf{p}-\mathbf{p}')\delta^{r-2}(\mathbf{q}-\mathbf{q}')\delta(q^+-q'^+))\ U_g(\varepsilon(s,\ p\bullet q/M)) \quad (5.41)$$
$$\{d_{iBT\gamma_R}{}^+(\ s,\ p,\ q,\ s_1,\ s_2),\ d_{jBT\gamma'_R}{}^{++}(\ s',\ p',\ q',\ s_1',\ s_2')\} =$$
$$= -(1 - \delta_{ij})\delta_{\gamma\gamma'}\delta_{s,s'}\delta_{s_1,s_1'}\delta_{s_2,s_2'}\ \delta^{r-1}(\mathbf{p}-\mathbf{p}')\delta^{r-2}(\mathbf{q}-\mathbf{q}')\delta(q^+-q'^+))\ U_g(\varepsilon(s,\ p\bullet q/M)) \quad (5.42)$$

where $N_R{}^+(p,\ q) = N_L{}^+(p,\ q)$, and where the other right-handed anti-commutation relations are zero. The spinors are given by eqs. 3.80 and 3.81, and Appendix 3-A.

5.6 Interior Particle Operators

We define the interior operators (tachyonic and gambol) with

$$b_{i\gamma L}{}^+{}_g(s,\ q,\ s_2) = \Sigma_{s_1} (\int d^{r-1}p\ N_L{}^{+'}(p,\ q)^{\frac{1}{2}}\ b_{i\gamma L}{}^+(s,\ p,\ q,\ s_1,\ s_2) \quad (5.43)$$
$$b_{i\gamma L}{}^+{}_g(s,\ q,\ s_2)^\dagger = \Sigma_{s_1} (\int d^{r-1}p\ N_L{}^{+'}(p,q)^{\frac{1}{2}}\ b_{i\gamma L}{}^+(s,\ p,\ q,\ s_1,\ s_2)^\dagger$$
$$b_{i\gamma R}{}^+{}_g(s,\ q,\ s_2) = \Sigma_{s_1} (\int d^{r-1}p\ N_R{}^{+'}(p,q)^{\frac{1}{2}}\ b_{i\gamma R}{}^+(s,\ p,\ q,\ s_1,\ s_2)$$
$$b_{i\gamma R}{}^+{}_g(s,\ q,\ s_2)^\dagger = \Sigma_{s_1} (\int d^{r-1}pN_R{}^{+'}(p,q)^{\frac{1}{2}}\ b_{i\gamma R}{}^+(s,\ p,\ q,\ s_1,\ s_2)^\dagger$$

---

[50] To obtain a non-gambol equivalent interior we set $U_g = 1$ and omit s and s' insertions.

where s is the fractionation and $N'(p,q) = (2\pi)^{r-1}N(p,q)$. Then we define the gambol field operator anti-commutators with the anti-commutation relations:

$$\{b_{i\gamma Lg}(s, q, s_2), b_{i\gamma' Lg}(s', q', s_2')^{\dagger}\} =$$
$$= (1 - \delta_{ij})\delta_{\gamma,\gamma'}\delta_{s,s'}\delta_{s_2,s_2'}2^{-1}[C^-R^+]_{ab}\delta(q^- - q'^-)\delta^{r-2}(q - q')U_g(\epsilon(s,q))$$

and

$$\{d^{1/s}{}_{g\gamma i}(s, q, s_2), d^{1/s'}{}_{g\gamma' j}(s', q', s_2')^{\dagger}\} =$$
$$= (1 - \delta_{ij})\delta_{s,s'}\delta_{s_2,s_2'}\delta_{\gamma,\gamma'}2^{-1}[C^-R^+]_{ab}\delta(q^- - q'^-)\delta^{r-2}(q - q')U_g(\epsilon(s,q))$$

$$(5.44)$$

where eq. 4.19 specifies $U_g(\epsilon(s,q))$ with $m = M$.

Eq. 5.43 has fractional integrations that utilize

$$[(\int dp^{r-1} N'(p,q))^{\frac{1}{2}}]^2 = \int dp^{r-1} N'(p,q) \qquad (5.45)$$

as in Riemann-Liouville integrals. Note γ and γ' are internal symmetry indices.

Thus particles and gambols have the same internal symmetries and spin. They differ in mass and momentum. The corresponding d type operators have similar anti-commutation relations. The other anti-commutators are zero:

$$\{b_{g\gamma i}(s, q, s_2)^{\dagger}, b_{g\gamma' j}(s', q', s_2')^{\dagger}\} = 0 \qquad (5.46)$$
$$\{d_{g\gamma i}(s, q, s_2)^{\dagger}, d_{g\gamma' j}(s', q', s_2')^{\dagger}\} = 0$$
$$\{b_{g\gamma i}(s, q, s_2), b_{g\gamma' j}(s', q', s_2')\} = 0$$
$$\{d_{g\gamma i}(s, q, s_2), d_{g\gamma' j}(s', q', s_2')\} = 0$$

and the anti-commutators of b and d type operators are zero as well.

The gambol creation/annihilation operators above can be used to define gambol fermion quantum fields.

5.7 External Particle Creation/Annihilation Operators

These operators may be defined using the composite creation/annihilation operators for particles of momentum p. We replace[51] the sum over s[52] with a sum over energies ϵ so as to take advantage of the Planckian normalization sum:[53]

$$b_{\gamma i}(p, s_2) = \int_0^{\infty} d\epsilon \, \Sigma_{s_1} (\int d^{r-1}q \, N'(p,q))^{\frac{1}{2}} b_{\gamma i}(s, p, q, s_1, s_2) \qquad (5.47)$$

$$b_{\gamma i}(p, s_2)^{\dagger} = \int_0^{\infty} d\epsilon \, \Sigma_{s_1} (\int d^{r-1}q \, N'(p,q))^{\frac{1}{2}} b_{\gamma i}(s, p, q, s_1, s_2)^{\dagger}$$

---

[51] We use s and ϵ interchangeably.

[52] The sum over s begins as a discrete sum over powers of 2 according to Cosmos Limos. We replace it with a continuous value for s, which makes the transition to an integral over ϵ possible.

[53] Using a sum over ϵ.

The external particle anti-commutation relations are[54]

$$\{b_{\gamma i}(p, s_2), b_{\gamma' j}(p', s_2')^\dagger\} = (1 - \delta_{ij})\,\delta_{s_2,s_2'}\,\delta_{\gamma\gamma'}\delta^{\,r-1}(\mathbf{p} - \mathbf{p'}) \tag{5.48}$$

We make the s (and ε) sums into integrations for analytic convenience just as the discrete sum over hv in the black body Planckian distribution derivation is similarly made continuous.

The other anti-commutation relations are:

$$\{d_{\gamma i}(p, s), d_{\gamma' j}(p', s')^\dagger\} = (1 - \delta_{ij})\,\delta_{s,s'}\,\delta_{\gamma\gamma'}\delta^{\,r-1}(\mathbf{p} - \mathbf{p'}) \tag{5.49}$$
$$\{b_{\gamma i}(p, s), b_{\gamma' j}(p', s')\} = 0$$
$$\{d_{\gamma i}(p, s), d_{\gamma' j}(p', s')\} = 0$$
$$\{b_{\gamma i}(p, s)^\dagger, b_{\gamma' j}(p', s')^\dagger\} = 0$$
$$\{d_{\gamma i}(p, s)^\dagger, d_{\gamma' j}(p', s')^\dagger\} = 0$$

and the anti-commutators of b and d type operators are zero.

The external particle[55] creation/annihilation operators above are those that appear in fermion quantum fields. Thus we may define an external *particle* quantum field with

$$\psi_{\gamma 1}(x) = \Sigma_p\,[b_{\gamma 1}(p, s)\,f_p(x) + d_{\gamma 1}(p, s)^\dagger f_p{}^*(x)] \tag{5.50}$$
$$\psi_{\gamma 2}(x) = \Sigma_p\,[b_{\gamma 2}(p, s)f_p(x) + d_{\gamma 2}(p, s)^\dagger f_p{}^*(x)]$$

with internal symmetry index γ.

---

[54] The s, s' and $\delta_{ss'}$ terms are equivalent to ε energy terms due to the 1:1 relation of s and ε.

[55] "External particle" means the particle as it appears as a particle without an interior part. The interior is masked by the external creation operator.

6. Duplex Fermion Bradyon-Tachyon ElectroWeak Theory

The purpose of this chapter is to probe more deeply into the internal ElectroWeak structure of fundamental particles. Recently we developed a new model of coupling constants and fermion internals.[56] Previously we developed a Gambol Theory[57] of particle interiors that appears quite successful. Now we more deeply explore the nature of ElectroWeak theory from a new viewpoint – a duplex fermion formulation. We are guided in this approach by the pairing of fundamental fermions: charged leptons paired on a one to one basis with neutral leptons and up-type quarks paired on a one to one basis with down-type quarks.[58]

We suggest these pairings reflect a new view of particle interactions. In particular we view the *interiors* of each ElectroWeak fermion pair as either a bradyon or a tachyon. This dichotomy naturally yields the pairs of fermions found in ElectroWeak Theory. In doing this, we note that we can separate a particle field into two parts: an external bradyon part whose coordinates govern the Physical dynamical motion of the particle and an internal part, which is either bradyon or tachyon. The interior part reveals itself through the pairings and the (broken) symmetry groups that they generate. The tachyonic nature within one of each fermion pair is masked and does not directly affect particle motion.

We will assume the lighter[59] of each pair's fermions has a tachyonic interior and the heavier of each pair has a bradyonic interior. Thus neutral leptons have tachyonic interiors and charged leptons have bradyonic interiors. Up-type quarks have bradyonic interiors and down-type quarks have tachyonic interiors. These lepton and quark mass regularities do not seem to have been understood in the past to the author's knowledge. Blaha (2024i) provides an analysis of fermion regularities.

The real external (LAB) part y coordinates are bradyonic. The types of the imaginary *internal* z coordinate wave function parts of the ElectroWeak pairs are:

$$e \qquad - \text{Bradyon}$$
$$\nu_e \qquad - \text{Tachyon}$$
$$u \text{ quark} - \text{Bradyon}$$
$$d \text{ quark} - \text{Tachyon}$$
$$s \text{ quark} - \text{Tachyon}$$
$$c \text{ quark} - \text{Bradyon}$$

---

[56] Blaha (2024i), *Particles and Universes of Cosmos Theory.*
[57] Blaha (2024a) *Cosmos-Universe-Particle-Gambol Theory.*
[58] This work provides an alternate view to our book on particle interiors, *Particles and Universes of Cosmos Theory* Blaha (2024i).
[59] The case of the u and d quarks is an exception since the u quark is lighter than the d quark—due to interaction effects.

$$b \text{ quark} - \text{Tachyon}$$
$$t \text{ quark} - \text{Bradyon}$$

and similarly for their other fermion generations.

The fermion wave functions generally differ. They may have bradyon or tachyon internal parts. As a result the Pauli matrices which are usually numeric must be generalized to have bradyon – tachyon projection operator factors. See eqs. 6.39 – 6.42 for their new form. The approach here, based on bradyon-Tachyon internal differences, is important for differentiating between fermions. It offers a fermion internal mechanism for understanding ElectroWeak interaction experimental data.

The leptons have additional real and imaginary coordinate systems for the SU(4) Strong Interaction behavior. See chapter 9 for these additional coordinate system parts.

In chapters 4 and 5 we developed two coordinate PseudoFermion quantum fields with one part (the y coordinates part) representing the external bradyonic fermion interface and the other part (the imaginary z coordinates part) representing the fermion interior. The possible exteriors are bradyonic. The possible interiors are bradyonic and tachyonic. We use that formalism here.

This design is implemented using PseudoFermion, PseudoQuantum fermion fields. The design deepens the pairing of the Lorentz group $SO(1,3)^+$ with $SU(2)\otimes U(1)$ seen in the Cosmos Theory dimension array for our universe.

Our approach in this chapter and the following chapters is to create sets of coordinate systems that combine interactions:[60]

1. $\text{ElectroWeak}\otimes SO(1, 3)^+$ is effectively combined using y coordinates as the real Lab coordinate system and z coordinates as an imaginary coordinate system. Together they make the complex coordinates $y + iz$. As a result we have bradyon and tachyon sectors.

2. For Strong Interaction SU(4) we use an additional pair of coordinate systems u and v that combine to make a complex coordinate system $u + iv$. This coordinate system is internal to fermions and not part of the Lab system. We find bradyon and tachyon sectors in a greater variety since these coordinates are private to the fermions. See chapter 9.

3. One can make a composite representation for quarks combining their ElectroWeak and SU(4) behavior. The result is the combined interactions $SO(1,3)^+\otimes SU(2)\otimes U(1)\otimes S(4)$ based on four coordinate systems: y, z, u and v to make $y + iz + uj + kv$. The combined coordinate systems may be put in a quaternion form[61] where each pair of 4 values is treated as a quaternion $Q^\mu = y^\mu + iz^\mu + ju^\mu + kv^\mu$.

---

[60] A variation of these possibilities was considered by the author in Blaha (2018e).
[61] Not done here.

In this chapter[62] we will develop the coordinate space form of the ElectroWeak interactions based on complex coordinates $y^\mu + iz^\mu$. We view the real-valued y coordinate system as the coordinate system (space), within which we reside and experience. We view the imaginary z coordinate system as an internal coordinate system that resides within fundamental fermions and bosons. The internal coordinate system may or may not be tachyonic since it is masked by the external y coordinate system. As a result the z coordinate system, when it is viewed as imaginary (superluminal), does not affect particles except by providing an ElectroWeak framework. It does not directly generate superluminal (faster than light) motions in the y coordinate system of our experience. The collapse of fermions to the vacuum is avoided.

ElectroWeak Doublets

We begin by generalizing the free Dirac equation to a 2×2 matrix of Dirac-like equations. This matrix equation is applied to a column of quantum fields forming a doublet representing ElectroWeak doublets. One type of ElectroWeak fermion quantum field $\psi(y, z)$ has a bradyon external part and a bradyon internal part. The other type of fermion $\psi_{T_z}(y, z)$ has a *bradyon external part* and a tachyon internal part.[63] A doublet column of these new types of fermion fields is represented by:

$$\psi_{BB}(y, z) \quad \text{or} \quad \psi_{BT}(y, z) \tag{6.1}$$

where B indicates a Bradyon field part and T represents a Tachyon field part. We will identify these doublets as representing ElectroWeak doublets. A charged fermion has a bradyon interior using $\psi_{BB}(y, z)$. Neutral fermions have tachyon interior using $\psi_{BT}(y, z)$.

SU(4) Quadruplets

In chapter 9 we consider the SU(4) Strong Interaction (that is later broken to SU(3) and U(1) sectors) for octets of fermions. The octets consist of fields with bradyonic and tachyonic parts. In this case we have an external bradyon part. One interior part is grouped with the external part. It can be either bradyonic or tachyonic. Another pair of parts can each be either bradyonic or tachyonic. The variations in the parts gives eight types of fermion fields. (See chapter 9.) We identify these octets as SU(4) fundamental representation fermion fields.[64] The SU(4) fundamental representation has four complex dimensions (eight real-valued) dimensions.

ElectroWeak Doublets Plan

We now turn to consider transformations in the imaginary coordinates z between bradyon and tachyon fields with a view to generating ElectroWeak SU(2)⊗U(1).

---

[62] This chapter is a variation on chapters in Blaha (2007b) and (2018e).
[63] The z subscript indicates a z coordinate system.
[64] SU(4) is dynamically broken to SU(3) quarks and a lepton U(1): SU(3)⊗U(1).

The plan of the chapter is:

1. Develop the form of a two coordinate system fermion field containing bradyon and tachyon parts.
2. Set up equations of motion for a fermion fields doublet with bradyon and tachyon parts.
3. Verify Lorentz covariance.
4. Develop a doublet implementation for ElectroWeak – SL(2, **C**).

Transformations between Sublight and Superluminal

6.1 Transformations between Sublight and Superluminal Coordinates

We begin with transformations between bradyon and tachyon fields using Left-handed boosts. A Left-handed boost has the form

$$\Lambda_L(\omega, \mathbf{u}) = \Lambda(\omega + i\pi/2, \mathbf{u}) = \exp[i\omega_L \hat{\mathbf{u}} \cdot \mathbf{K}] \qquad (3.23)$$

where **K** is the boost vector (seen below)

$$\omega_L = \omega + i\pi/2 \qquad (3.23a)$$

and **u** is the unit vector

$$\mathbf{u} = \mathbf{v}/|\mathbf{v}| \qquad (3.23b)$$

with **v** being the relative velocity. Thus

$$\cosh(\omega_L) = i\,\sinh(\omega) = -\gamma = i\gamma_s \qquad (3.24)$$
$$\sinh(\omega_L) = i\,\cosh(\omega) = -\beta\gamma = i\beta\gamma_s$$

with, $\beta = v > 1$, $\gamma_s = (\beta^2 - 1)^{-\frac{1}{2}}$, and $\omega \geq 0$. The rapidity w is

$$w = \operatorname{artanh}(\omega_L) = \beta \qquad (3.24a)$$

Thus

$$\sinh(\omega) = \gamma_s \qquad (3.25)$$
$$\cosh(\omega) = \beta\gamma_s$$
$$w^{-1} = \operatorname{artanh}(\omega) = \beta^{-1} \qquad (3.24a)$$

A (left-handed[65]) Lorentz transformation is

$$\Lambda_L(\omega, \mathbf{u} = (1,0,0)) = \begin{bmatrix} i\gamma_s & -i\beta\gamma_s & 0 & 0 \\ -i\beta\gamma_s & i\gamma_s & 0 & 0 \\ 0 & 0 & 1 & 0 \\ 0 & 0 & 0 & 1 \end{bmatrix} \qquad (6.2)$$

It implements the coordinate transformation (Fig. 6.1):

---

[65] The right-handed Lorentz transformation case is analogous.

$$X' = \Lambda_L(\omega, \mathbf{u} = (1,0,0))X$$

or

$$
\begin{aligned}
t' &= i\gamma_s(t - \beta x)\\
x' &= i\gamma_s(x - \beta t)\\
y' &= y\\
z' &= z
\end{aligned}
\tag{6.3}
$$

Thus a Left-handed boost transforms four real-valued coordinates into imaginary counterparts. The metric tensor for the imaginary coordinates takes diag $g^{\mu\nu} = (1, -1, -1, -1)$ into diag $g^{\mu\nu} = (-1, 1, 1, 1)$.

The corresponding spinor transformation for a fermion field is:

$$
\begin{aligned}
S_L(\Lambda_L(\omega, \mathbf{u})) &= \exp(-i\omega_L\sigma_{0i}v_i/(2|\mathbf{v}|)) = \exp(-\omega_L\gamma^0\boldsymbol{\gamma}\cdot\mathbf{v}/(2|\mathbf{v}|))\\
&= \cosh(\omega_L/2)I + \sinh(\omega_L/2)\gamma^0\boldsymbol{\gamma}\cdot\mathbf{p}/|\mathbf{p}|
\end{aligned}
\tag{3.26}
$$

The inverse transformation is

$$
\begin{aligned}
S_L^{-1}(\Lambda_L(\omega, \mathbf{u})) &= \gamma^2\gamma^0\kappa^{-1}S_L^\dagger\kappa\gamma^0\gamma^2 = \gamma^2\gamma^0 S_L{}^\tau\gamma^0\gamma^2 = \exp(\omega_L\gamma^0\boldsymbol{\gamma}\cdot\mathbf{v}/(2|\mathbf{v}|))\\
&= \cosh(\omega_L/2)I - \sinh(\omega_L/2)\gamma^0\boldsymbol{\gamma}\cdot\mathbf{p}/|\mathbf{p}|
\end{aligned}
\tag{3.27}
$$

where the superscript τ denotes the transpose and κ is the complex conjugation operator (that also appears in the time-reversal operator). The spinor boost vector is

$$\mathbf{K} = i\gamma^0\boldsymbol{\gamma}/2 \tag{3.27a}$$

Note that S_L is not unitary just as the equivalent spinor Lorentz transformation $S(\Lambda(v))$ is not unitary.

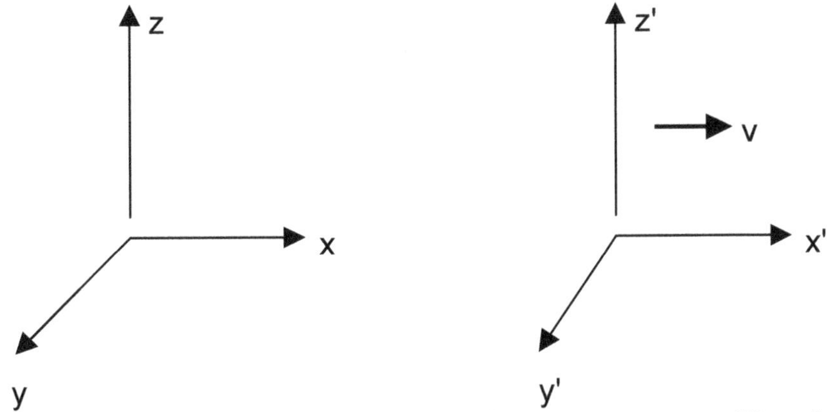

Figure 6.1. Depiction of two coordinate systems. The "primed" coordinate system is moving with velocity v in the positive x direction with respect to the "unprimed" coordinate system. We choose parallel axes for convenience.

6.1.1 The Boost Operators K

There are three boost operators K_1, K_2 and K_3 with the form:

$$K_1 \;\dot= \; \begin{bmatrix} 0 & 1 & 0 & 0 \\ 1 & 0 & 0 & 0 \\ 0 & 0 & 0 & 0 \\ 0 & 0 & 0 & 0 \end{bmatrix} \tag{6.4}$$

$$K_2 \;=\; \begin{bmatrix} 0 & 0 & 1 & 0 \\ 0 & 0 & 0 & 0 \\ 1 & 0 & 0 & 0 \\ 0 & 0 & 0 & 0 \end{bmatrix} \tag{6.5}$$

$$K_3 = \begin{bmatrix} 0 & 0 & 0 & 1 \\ 0 & 0 & 0 & 0 \\ 0 & 0 & 0 & 0 \\ 1 & 0 & 0 & 0 \end{bmatrix} \tag{6.6}$$

where

$$[K_i, K_j] = 0$$

Bradyon – Tachyon Transformations

6.2 PseudoFermion Bradyon – Tachyon Field Transformation

We now consider tachyonic transformations of the interior z coordinate part of a PseudoFermion field. Eqs. 3.26 – 3.27 for the z coordinate system of a free bradyon field transforms it into a tachyon field, and vice versa:[66]

$$S_{zL}(\Lambda_L(\omega, \mathbf{u}))\psi(y, z) \rightarrow \psi_{T_z}'(y, z') \tag{6.7}$$
$$S_{zL}(\Lambda_L(\omega, \mathbf{u}))\psi_{T_z}(y, z) \rightarrow \psi'(y, z')$$

where $S_{zL}(\Lambda_L(\omega, \mathbf{u})) = S_L(\Lambda_L(\omega, \mathbf{u}))$ as in section 3.4.3. Note the y coordinate system is unchanged by the transformation. Also, note the appearance of a γ^5:

$$S_{zL}(\Lambda_{zL}(\omega, \mathbf{u}))(\gamma^\mu\partial/\partial z^\mu - M)S_{zL}^{-1}(\Lambda_{zL}(\omega, \mathbf{u})) = (i\gamma^\mu\partial/\partial z'^\mu - M) \tag{6.8}$$
$$S_{zL}(\Lambda_{zL}(\omega, \mathbf{u}))\gamma^5(i\gamma^\mu\partial/\partial z^\mu - M)\gamma^5 S_{zL}^{-1}(\Lambda_{zL}(\omega, \mathbf{u})) = (\gamma^\mu\partial/\partial z'^\mu - M)$$

where

$$z'^\mu = i\Lambda_{zL}{}^\mu{}_\nu(\omega, \mathbf{u})z^\nu \tag{6.9}$$
$$\partial/\partial z'^\mu = -i\Lambda_{zL}{}^\nu{}_\mu(\omega, \mathbf{u})\partial/\partial z^\nu$$

with

---

[66] Note we insert a z subscript to indicate the transformation is for the z coordinates.

$$z' = E(\mathbf{v})z = i\Lambda_{zL}(\mathbf{v})z$$

The above equations imply the transformations of equations of motion is:

$$S_{zL}(\Lambda_{zL}(\omega, \mathbf{u}))(\gamma^{\mu}\partial/\partial z^{\mu} - M)\psi_{T_z}(y,z) = (i\gamma^{\mu}\partial/\partial z'^{\mu} - M)S_{zL}(\Lambda_{zL}(\omega, \mathbf{u}))\psi_{T_z}(y,z)$$
$$= (i\gamma^{\mu}\partial/\partial z'^{\mu} - M)\psi'(y,z') \qquad (6.10)$$

and

$$S_{zL}(\Lambda_{zL}(\omega, \mathbf{u}))\gamma^5(i\gamma^{\mu}\partial/\partial z^{\mu} - M)\psi(y,z) = (\gamma^{\mu}\partial/\partial z'^{\mu} - M)S_{zL}(\Lambda_{zL}(\omega, \mathbf{u}))\gamma^5\psi(y,z)$$
$$= (\gamma^{\mu}\partial/\partial z'^{\mu} - M)\psi_{T_z}'(y,z') \qquad (6.11)$$

where

$$\psi'(y,z') = S_{zL}(\Lambda_{zL}(\omega, \mathbf{u}))\psi_{T_z}(x) \qquad (6.12)$$
$$\psi_{T_z}'(y,z') = S_{zL}(\Lambda_{zL}(\omega, \mathbf{u}))\gamma^5\psi(y,z) \qquad (6.13)$$

<u>Sublight – Superluminal Doublet Dynamic Equation</u>

6.3 Doublet Extended Field Equation

We will now consider the issue of generalizing the field equation so that the extended equation is covariant under both Lorentz transformations and Left-handed Complex Lorentz transformations.

The only obvious method to obtain an extended field equation that is covariant under Complex Lorentz transformations is to define an 8×8 matrix generalization. Let

The equations of motion from eqs. 4.2 and 5.2 are[67]

$$[iM^{-1}\gamma_y^{\mu}\partial/\partial y^{\mu}\,\gamma_z^{\nu}\partial/\partial z^{\nu} - M]\psi_{Tz} = 0 \qquad (5.2a)$$
$$[-M^{-1}\gamma_y^{\mu}\partial/\partial y^{\mu}\,\gamma_z^{\nu}\partial/\partial z^{\nu} - M]\psi = 0 \qquad (4.2a)$$

The subsidiary equations of motion for the bradyon field parts are:

$$B_z\psi = [i\gamma_z^{\mu}\partial/\partial z^{\mu} - M]\psi = 0 \qquad (4.2c)$$
$$T_z\psi_T = [\gamma_z^{\mu}\partial/\partial z^{\mu} - M]\psi_T = 0 \qquad (5.2c)$$

The doublet extended field equation operators are

$$đ(y, z) = \begin{bmatrix} T_z & 0 \\ 0 & B_z \end{bmatrix} \qquad (6.14)$$

They define an 8×8 matrix operator with the 4×4 operator matrix elements B and T. Let

---

[67] We use conventional Quantum Field theory in this chapter for convenience. It may be directly generalized to PseudoQuantum Field Theory.

$$\Psi(y,z) \;=\; \begin{bmatrix} \psi_T(y,z) \\[20pt] \psi(y,z) \end{bmatrix} \tag{6.15}$$

be an 8 component column vector. Then the extended free field equation is

$$đ(y,\,z)\Psi(y,z) = 0 \tag{6.16}$$

We now define the 8×8 Left-handed Complex Lorentz tachyonic transformation

$$S_{zL8}(\Lambda_{zL}(\omega,\,\mathbf{u})) \;=\; \begin{bmatrix} 0 & S_{zL}(\Lambda_{zL}(v))\gamma^5 \\[20pt] S_{zL}(\Lambda_{zL}(v)) & 0 \end{bmatrix} \tag{6.17}$$

with inverse transformation

$$S_{zL8}^{\;-1}(\Lambda_{zL}(v)) \;=\; \begin{bmatrix} 0 & S_{zL}^{\;-1}(\Lambda_{zL}(v)) \\[20pt] \gamma^5 S_{zL}^{\;-1}(\Lambda_{zL}(v)) & 0 \end{bmatrix} \tag{6.18}$$

Note: we use the notations $S_{zL}(\Lambda_{zL}(v))$ and $S_{zL}(\Lambda_{zL}(\omega,\,\mathbf{u}))$ interchangeably. Applying S_{zL8} yields

$$0 = S_{zL8}(\Lambda_L(v))đ(y,\,z)\Psi(y,z) = đ(y,\,z')\Psi'(y,z') \tag{6.19}$$

where

$$\Psi'(y,z') \;=\; \begin{bmatrix} S_{zL}\gamma^5\psi(y,z) \\[20pt] S_{zL}\psi_{T_z}(y,z) \end{bmatrix} \;=\; \begin{bmatrix} \psi_{T_z}'(y,z') \\[20pt] \psi'(y,z') \end{bmatrix} \tag{6.20}$$

The extended equation "exchanges" the bradyon and tachyon fields in the z coordinates part.

It is easy to show that the extended field equation above is also covariant under conventional Lorentz transformations in the 8×8 representation:

$$S_{z8}(\Lambda_z (v)) = \begin{bmatrix} S_z(\Lambda_z(v)) & 0 \\ 0 & S_z(\Lambda_z(v)) \end{bmatrix} \tag{6.21}$$

with inverse

$$S_{z8}^{-1}(\Lambda(v)) = \begin{bmatrix} S_z^{-1}(\Lambda_z (v)) & 0 \\ 0 & S_z^{-1}(\Lambda_z (v)) \end{bmatrix} \tag{6.22}$$

and non-diagonal Lorentz transformations:

$$S_{z8A}(\Lambda_z (v)) = \begin{bmatrix} 0 & S_z(\Lambda_z (v)) \\ S_z(\Lambda_z(v)) & 0 \end{bmatrix} \tag{6.23}$$

with inverse transformation

$$S_{z8A}^{-1}(\Lambda(v)) = \begin{bmatrix} 0 & S_z^{-1}(\Lambda_z(v)) \\ S_z^{-1}(\Lambda_z(v)) & 0 \end{bmatrix} \tag{6.24}$$

Under a conventional Lorentz transformation we find

$$0 = S_{z8}(\Lambda_z(v))đ(y, z)\Psi(y,z) = đ(y, z')\Psi'(y,z') \tag{6.25}$$
$$0 = S_{z8A}(\Lambda_z(v))đ(y, z)\Psi(y,z) = đ(y, z')\Psi'(y,z')$$

The Lagrangian density that corresponds to our 8-dimensional construction is

$$\mathcal{L}_8 = \overline{\Psi}(y,z)đ(y,z)\Psi(y,z) \tag{6.26}$$

using eq. 6.14.where

$$\overline{\Psi}(x) = \Psi^\dagger \Gamma^0 \tag{6.27}$$

and using eq. 5.2a

$$\Gamma^0 = \begin{bmatrix} i\,\gamma_y^{\ 0}\gamma_z^{\ 0}\gamma_z^{\ 5} & 0 \\ 0 & \gamma_y^{\ 0} \end{bmatrix} \tag{6.28}$$

The action I:

$$I = \int d^4y \, d^4z \, \mathcal{L}_8 \tag{6.29}$$

is invariant under generalized Lorentz transformations S_{z8} and S_{z8A}.

Minimal Tachyonic Transformations

6.4 Minimal Tachyonic Transformations

The most minimal tachyonic transformation simply maps z to z'. It allows us to simplify the effect of the eq. 6.17 and 6.18 tachyonic transformation to

$$\Psi'(y,z') = \begin{bmatrix} S_{zL}\gamma^5\psi(y,z) \\[2ex] S_{zL}\psi_{T_z}(y,z) \end{bmatrix} = \begin{bmatrix} \psi_{T_z}'(y,\,iz) \\[2ex] \psi'(y,\,iz) \end{bmatrix} \tag{6.30}$$

based on eqs. 3.23- 3.24 and 6.2 - 6.4 above. The coordinates z are transformed to

$$\begin{aligned} z'^0 &= iz^0 \\ z'^z &= iz^z \\ z'^x &= z^x \\ z'^y &= z^y \end{aligned} \tag{6.31}$$

The minimalist field transformation takes

$$S_L(\Lambda_L(\omega,\,\mathbf{u})) = \exp(-i\omega_L\sigma_{0i}v_i/(2|\mathbf{v}|)) = \exp(-\omega_L\gamma^0\boldsymbol{\gamma}\cdot\mathbf{v}/(2|\mathbf{v}|))$$
$$= i\sinh(\omega/2)I + i\cosh(\omega/2)\gamma^0\boldsymbol{\gamma}\cdot\mathbf{u} \tag{3.26}$$

to

$$S_L(\Lambda_L(\omega,\,\mathbf{u})) = i\gamma^0\boldsymbol{\gamma}\cdot\mathbf{u} = i\gamma^0\gamma_3 \tag{6.32}$$

where $\omega = 0$ and $u_z = \gamma_3$.

The inverse transformation is

$$S_L^{-1}(\Lambda_L(\omega,\,\mathbf{u})) = \gamma^2\gamma^0\kappa^{-1}S_L^{\dagger}\kappa\gamma^0\gamma^2 = \gamma^2\gamma^0 S_L{}^{\tau}\gamma^0\gamma^2 = \exp(\omega_L\gamma^0\boldsymbol{\gamma}\cdot\mathbf{v}/(2|\mathbf{v}|))$$
$$= i\sinh(\omega/2)I - i\cosh(\omega/2)\gamma^0\boldsymbol{\gamma}\cdot\mathbf{u} \tag{3.27}$$

and

$$S_L^{-1}(\Lambda_L(\omega,\,\mathbf{u})) = -i\,\gamma^0\boldsymbol{\gamma}\cdot\mathbf{u} = -i\gamma^0\gamma_3 \tag{6.33}$$

where $\omega = 0$ and $u_z = \gamma_3$. Note:the squares of the minimal transformations are:

$$S_L(\Lambda_L(0,\,\mathbf{u}))^2 = -I \tag{6.33a}$$
$$S_L^{-1}(\Lambda_L(0,\,\mathbf{u}))^2 = -I \tag{6.33b}$$

Then eq. 6.30 becomes

$$\Psi'(y,iz) \;=\; \begin{bmatrix} S_{zL}\gamma^5\psi(y,z) \\[2mm] S_{zL}\psi_{T_z}(y,z) \end{bmatrix} \;=\; \begin{bmatrix} i\gamma^0\gamma_3\gamma^5\psi(y,\,iz) \\[2mm] i\gamma^0\gamma_3\psi(y,\,iz) \end{bmatrix} \tag{6.34}$$

again using the unit vector $\mathbf{u} = \check{\mathbf{z}}$, which we chose, based on the light front formalism in chapter 3:

$$\gamma^\pm = (\gamma^0 \pm \gamma^3)/\sqrt{2} \tag{3.57}$$

with transverse matrices γ^1 and γ^2 defined as usual. Note the useful identity:

$$\gamma^{\pm\,2} = 0$$

Light Front Tachyon Fields

We defined "+" and "–" tachyon fields with the projection operators:

$$R^\pm = \tfrac{1}{2}(I \pm \gamma^0\gamma^3) \tag{3.58}$$

giving

Left-handed, $\pm$ light-front fields: $\qquad \psi_{TL}{}^\pm = R^\pm C^-\psi_T$

$$\tag{3.59}$$

Right-handed, $\pm$ light-front fields: $\qquad \psi_{TR}{}^\pm = R^\pm C^+\psi_T$

Bradyon – Tachyon ElectroWeak Theory

6.5 Bradyon-Tachyon ElectroWeak Theory

The Lagrangian density of eq. 6.26 may be extended to include the gauge vector fields of ElectroWeak SU(2)⊗U(1) theory. Starting from the non-Weinberg rotated covariant derivative:

$$D^\mu = \partial^\mu + igt\cdot W^\mu + ig't_0 W_0{}^\mu \tag{6.35}$$

we generalize to the two coordinate sets y and z with exterior coordinates y being bradyon always and interior coordinates z being either bradyonic or tachyonic as the case may be. We do not introduce left and right handed fields as yet. Thus the free Lagrangian:[68]

$$\mathcal{L} = \overline{\psi}_{2T}{}^{\alpha\beta}[iM^{-1}\gamma_{y\alpha\kappa}{}^\mu\partial/\partial y^\mu\gamma_{z\beta\lambda}{}^\nu\partial/\partial z^\nu - M]\psi_{1T}{}^{\kappa\lambda} +$$
$$+\;\overline{\psi}_{1T}{}^{\alpha\beta}[iM^{-1}\gamma_{y\alpha\kappa}{}^\mu\partial/\partial y^\mu\,\gamma_{z\beta\lambda}{}^\nu\partial/\partial z^\nu - M]\psi_{2T}{}^{\kappa\lambda} \tag{5.2}$$

---

[68] Note PseudoQuantum fields.

where γ_y^{μ} and γ_z^{μ} are Dirac matrices for the y and z coordinates respectively, and y and z are coordinates in r dimension space-times, M is the mass, and

$$\overline{\psi}_{kT\alpha\beta} = \psi_{kT}^{\ \kappa\lambda\dagger}\gamma_y^{\ 0}_{\ \kappa\alpha}\ i\gamma_z^{\ 0}\gamma_z^{\ 5}_{\ \lambda\beta} \tag{5.2a}$$

for i, k = 1, 2 where the subscripts y and y indicate Dirac spinors associated with the y and z coordinates.

We now introduce the gauge fields using eq. 6.35 above while restricting the gauging to the y coordinates part so as to recover the conventional equations in the y coordinates sector:

$$[iM^{-1}\gamma_y^{\ \mu}\ D_y^{\ \mu}\ \gamma_z^{\ \nu}\partial/\partial z^{\nu} - M]\psi_{jT} = 0 \tag{6.36}$$

with j = 1, 2 using eq. 6.35.

The above development of transformations between bradyon and tachyon fields can be used to replace the SU(2)⊗U(1) matrices with numeric coordinates with minimalistic operators from eqs.6.32 and 6.35 which we denote as $t(S_L)$ and $t_0(S_L)$

$$D^{\mu}(y) = \partial^{\mu} + ig\mathbf{t}(S_{zL})W^{\mu}(y) + ig't_0(S_{zL})W_0^{\ \mu}(y) \tag{6.37}$$
$$D^{i}(y) = \partial^{i} + ig\mathbf{t}(S_{zL})W^{i}(y) \tag{6.37a}$$
$$D^{0}(y) = \partial^{0} + ig't_0(S_{zL})W_0^{\ 0}(y) \tag{6.37b}$$

The matrices have a Pauli matrix form with the numbers in the Pauli matrices replaced with minimalist operators S_L and S_L^{-1} from eqs. 6.32 and 6.33. These operators are numeric and thus are similar to coupling constants. Thus the y subsidiary equation (eq. 5.2c) becomes

$$[i\gamma_y^{\ \mu}D^{\mu}(y) - M]\psi(y,z) = [i\gamma_y^{\ \mu}D^{\mu}(y) - M]\psi_T(y,z) = 0 \tag{6.38}$$

The z coordinates equation is independent of the gauge fields.

6.5.1 Pauli-type S_{Lz} Dependent Matrices

The Pauli matrices that appear in the ElectroWeak Lagrangian must be enhanced due to the different forms of the fermion fields: bradyon charged leptons and tachyon neutral leptons. Quark ElectroWeak theory embodies the same need for enhanced Pauli matrices in its Lagrangian.

The enhanced Pauli matrices depend on transformations between the bradyon and tachyon z parts of fields. They have the form

$$t_- = \tfrac{1}{2}\begin{bmatrix} 0 & S_{zL}\gamma^5 \\ 0 & 0 \end{bmatrix} \tag{6.39}$$

$$= \tfrac{1}{2} \begin{bmatrix} 0 & i\gamma_z^{\ 0}\gamma_{z3}\gamma_z^{\ 5} \\ 0 & 0 \end{bmatrix}$$

$$t_+ = \tfrac{1}{2} \begin{bmatrix} 0 & 0 \\ S_{zL} & 0 \end{bmatrix} \tag{6.40}$$

$$= \tfrac{1}{2} \begin{bmatrix} 0 & 0 \\ i\gamma_z^{\ 0}\gamma_{z3} & 0 \end{bmatrix}$$

$$t_3 = \tfrac{1}{2} \begin{bmatrix} -i\gamma_z^{\ 0}\gamma_{z3}\gamma_z^{\ 5} & 0 \\ 0 & i\gamma_z^{\ 0}\gamma_{z3} \end{bmatrix} \tag{6.41}$$

$$\equiv \tfrac{1}{2} \begin{bmatrix} -1 & 0 \\ 0 & 1 \end{bmatrix}$$

$$t_0 = \tfrac{1}{2} \begin{bmatrix} 1 & 0 \\ 0 & 1 \end{bmatrix} \tag{6.42}$$

The electric charge is

$$Q = t_0 + t_3 \tag{6.43}$$

and the interaction terms become

$$t(S_{zL}) \cdot W^i(y) \rightarrow t_- W^+ + t_+ W^- + t_3 W^3 \tag{6.43a}$$

$$(S_{zL}) W_0^{\ 0}(y) \rightarrow t_0 W^0 \tag{6.43b}$$

6.5.2 Creation and Annihilation Operators

Creation and annihilation operators may be defined in a manner similar to the development in chapter 5. If "handed" and light front operators are defined as in eqs. 5.35 and 5.36 we obtain the operator anti-commutators:

$$\{b_{iT\gamma L}{}^{+}(p, q, s_1, s_2), b_{jT\gamma' L}{}^{++}(p', q', s_1', s_2')\} =$$
$$= (1 - \delta_{ij})\delta_{\gamma\gamma'}\,\delta_{s_1,s_1'}\delta_{s_2,s_2'}\,\delta^{r-1}(\mathbf{p} - \mathbf{p}')\delta^{r-2}(\mathbf{q} - \mathbf{q}')\delta(q^+ - q'^+)) \qquad (6.44)$$

$$\{d_{iT\gamma L}{}^{+}(p, q, s_1, s_2), d_{jT\gamma' L}{}^{++}(p', q', s_1', s_2')\} =$$
$$= (1 - \delta_{ij})\delta_{\gamma\gamma'}\delta_{s_1,s_1'}\delta_{s_2,s_2'}\,\delta^{r-1}(\mathbf{p} - \mathbf{p}')\delta^{r-2}(\mathbf{q} - \mathbf{q}')\delta(q^+ - q'^+)) \qquad (6.45)$$

where $i, j = 1, 2$ are PseudoQuantum indices; γ and γ' are symmetry indices; and s_1, s_1', s_2, and s_2' are spins.

One may choose equal spins: $s_1 = s_1'$ and $s_2 = s_2'$ to support the entanglement mechanism discussed in item 5 in the next subsection.

Bradyon – Tachyon ElectroWeak Lagrangian

6.5.3 Lagrangian Formulation

From eq. 5.31 we find the form of the fermion Lagrangian using eq. 5.27:

$$\Psi_{iTL\alpha}{}^{\lambda} = i(\gamma_y{}^0\gamma_y)_{\alpha\kappa}{}^{\mu}\partial/\partial y^{\mu}\psi_{iTL}{}^{\kappa\lambda}$$
$$\Psi_{iTR\alpha}{}^{\lambda} = i(\gamma_y{}^0\gamma_y)_{\alpha\kappa}{}^{\mu}\partial/\partial y^{\mu}\psi_{iTR}{}^{\kappa\lambda}$$

which we express in PseudoQuantum form using eq. 6.37:

$$\Psi_{iTL} = i\gamma_y{}^0\gamma_y{}^k D_k \psi_{iTL} \qquad (6.46)$$
$$\Psi_{iTR} = i\gamma_y{}^0\gamma_y{}^k D_k \psi_{iTR}$$
$$\Psi_{iTL}{}^0 = i\gamma_y{}^0\gamma_y{}^0 D_0\psi_{iTL}$$
$$\Psi_{iTR}{}^0 = i\gamma_y{}^0\gamma_y{}^0 D_0\psi_{iTR}$$

$$\mathcal{L} = i\psi_{2TL}{}^{\dagger}\gamma_y{}^0\gamma_z{}^0\gamma_z{}^v\partial/\partial z^v\Psi_{1TL} + i\psi_{2TL}{}^{\dagger}\gamma_y{}^0\gamma_z{}^0\gamma_z{}^v\partial/\partial z^v\Psi_{1TL}{}^0 - iM\psi_{2TR}{}^{\dagger}\gamma_y{}^0\gamma_z{}^0\psi_{1TL} -$$
$$- i\psi_{2TR}{}^{\dagger}\gamma_y{}^0\gamma_z{}^0\gamma_z{}^v\partial/\partial z^v\Psi_{1TR} - i\psi_{2TR}{}^{\dagger}\gamma_y{}^0\gamma_z{}^0\gamma_z{}^v\partial/\partial z^v\Psi_{1TR}{}^0 + iM\psi_{2TL}{}^{\dagger}\gamma_y{}^0\gamma_z{}^0\psi_{1TR} +$$
$$+ i\psi_{1TL}{}^{\dagger}\gamma_y{}^0\gamma_z{}^0\gamma_z{}^v\partial/\partial z^v\Psi_{2TL} + i\psi_{1TL}{}^{\dagger}\gamma_y{}^0\gamma_z{}^0\gamma_z{}^v\partial/\partial z^v\Psi_{2TL}{}^0 - iM\psi_{1TR}{}^{\dagger}\gamma_y{}^0\gamma_z{}^0\psi_{2TL} -$$
$$- i\psi_{1TR}{}^{\dagger}\gamma_y{}^0\gamma_z{}^0\gamma_z{}^v\partial/\partial z^v\Psi_{2TR} - i\psi_{1TR}{}^{\dagger}\gamma_y{}^0\gamma_z{}^0\gamma_z{}^v\partial/\partial z^v\Psi_{2TR}{}^0 + iM\psi_{1TL}{}^{\dagger}\gamma_y{}^0\gamma_z{}^0\psi_{2TR}$$
$$(6.47)$$

Eq. 6.47 may be extended to a handed and light front Lagrangian from eq. 5.32. In this case the $\gamma_z{}^0\gamma_{z3}$ factors in t_i "disappear" due to the light front operators

$$R^{\pm}\gamma_z{}^0\gamma_{z3} = \tfrac{1}{2}(I \pm \gamma_z{}^0\gamma_z{}^3) = -R^{\mp} \qquad (6.48)$$

6.5.4 Quantum Entanglement

The superluminal z coordinates part may be used to explain aspects of quantum entanglement in quantum mechanics and quantum field theory. If one requires the z coordinates spin to equal the y coordinates spin then a long range entanglement mechanism is possible.

If a multi-particle system is placed in a certain spin state and the state evolves to separate its parts, then the z parts of the particles can communicate "instantaneously" through their superluminal z parts. While the y parts of particles are localized, the z parts are not localized.

The z parts may thus extend throughout the universe and thereby support instantaneous communication of spins and possibly other particle features. The faster than light z parts can combine to support the functional solution for entanglement provided in Blaha (2018e).

6.5.5 Particle Interior Gambol Formulation

The z parts of particles can be easily modified to support a gambol implementation. In this implementation the gambol mass that is used in the Planckian distribution is m_g = M/s. The bradyon or tachyon nature of the interior part of the particle is assumed not to affect the distribution.

The creation and annihilation operators' anti-commutation relations in light front coordinates are:

$$\{b_{iT\gamma L}{}^{+}(s, p, q, s_1, s_2), b_{jT\gamma'L}{}^{++}(s', p', q', s_1', s_2')\} =$$
$$= (1 - \delta_{ij})\delta_{\gamma\gamma'}\,\delta_{s,s'}\delta_{s_1,s_1'}\delta_{s_2,s_2'}\,\delta^{r-1}(\mathbf{p \text{-} p'})\delta^{r-2}(\mathbf{q} - \mathbf{q'})\delta(q^{+} - q'^{+}))\, U_g(\varepsilon(s, p{\bullet}q/M)) \quad (5.35)$$

$$\{d_{iT\gamma L}{}^{+}(s, p, q, s_1, s_2), d_{jT\gamma'L}{}^{++}(s', p', q', s_1', s_2')\} =$$
$$= (1 - \delta_{ij})\delta_{\gamma\gamma'}\delta_{s,s'}\delta_{s_1,s_1'}\delta_{s_2,s_2'}\,\delta^{r-1}(\mathbf{p \text{-} p'})\delta^{r-2}(\mathbf{q} - \mathbf{q'})\delta(q^{+} - q'^{+}))\, U_g(\varepsilon(s, p{\bullet}q/M)) \quad (5.36)$$

where i, j = 1, 2 are PseudoQuantum indices; γ and γ' are symmetry indices; s and s' are gambol fractionations, p is the y coordinates momentum, q is the z coordinates momentum, and s_1, s_1', s_2, and s_2' are spins. U_g is a gambol Planckian distribution.

6.5.6 Impact of Bradyon-Tachyon ElectroWeak Theory

Bradyon-Tachyon ElectroWeak Theory has significant implications:

1. The theory provides a coordinate realization equivalent to SU(2)⊗U(1).

2. The theory implements a type of complex coordinate system where the y and z coordinates reflect a direct sum.

3. While the mass for both neutral and charged fermion is the same, namely M, the Higgs Mechanism can supply mass breaking to different masses.

4. The formulation provided in this chapter supports quark ElectroWeak Theory as well as lepton ElectroWeak Theory.

5. There have been successful efforts to combine the Strong Interactions (SU(3)) and ElectroWeak theory within other frameworks. The Bradyon-Tachyon ElectroWeak

Theory combines SU(2)⊗U(1) with SL(2, **C**) based on the form of Cosmos Theory for our universe. In this case, SU(2)⊗U(1) and SL(2, **C**) suggestively lie in the same subset of 8 dimensions in the Cosmos dimension array for symmetries. The Bradyon-Tachyon ElectroWeak Theory offers an alternative that exploits that *consanguinity*.

The Strong Interaction Bradyon-Tachyon formulation presented later in chapter 9 is also based on the 8 real dimensions in the SU(4) part of the Cosmos dimension array for symmetries.

6.6 Faster Than Light?

The appearance of tachyon fields for neutrinos would suggest that neutrinos move at faster than light and have other "troublesome" implications. However this is not true in the present case of PseudoFermion fields since the y coordinate system motion, which is the motion that experiment detects, is conventional bradyon motion. The faster than light motion, present in a tachyon interior part, is masked by the PseudoFermion formulation. Faster than light motion does not directly happen.

The tachyon internal structure of neutral fermions plays a role in determining the ElectroWeak theory structure's pairing of femions and symmetry groups.

7. Duplex Fermion Exterior Tachyon – Interior Bradyon Quantum Fields

In this chapter we specify *duplex fermion* quantum fields with an exterior tachyon and interior bradyon part by using the development of quantum fields with an exterior bradyon and interior tachyon parts in chapter 5:

$$\psi_{iTB\alpha\beta}(y, z) = \psi_{iBT\beta\alpha}(z, y) \qquad (7.1)$$

Eq. 7.1 enables us to use the development in chapter 5 to construct an exterior tachyon and interior bradyon quantum field.

8. Duplex Fermion Exterior Tachyon – Interior Tachyon Quantum Fields

We begin by defining a free *duplex fermion* PseudoFermion PseudoQuantum Lagrangian with two related quantum fields ψ_{1TT} and ψ_{2TT} that are functions of the two sets of coordinates in r space-time dimensions, y and z with the same metric tensor: $g_{\mu\nu y}$ = diag(1, –1, –1, –1) = $g_{\mu\nu z}$. The y coordinates are for an external LAB tachyon field. The z coordinates are also for a tachyon (internal) field. We implement what may be viewed as a form of complex space-time and define Dirac matrices correspondingly: $\gamma_{y\alpha\kappa}{}^{\mu}$ and $\gamma_{z\beta\lambda}{}^{\nu}$

$$\mathcal{L} = \overline{\psi}_{2TT}{}^{\alpha\beta}[-M^{-1}\gamma_{y\alpha\kappa}{}^{\mu}\partial/\partial y^{\mu}\gamma_{z\beta\lambda}{}^{\nu}\partial/\partial z^{\nu} + M]\psi_{1TT}{}^{\kappa\lambda} +$$
$$+ \overline{\psi}_{1TT}{}^{\alpha\beta}[-M^{-1}\gamma_{y\alpha\kappa}{}^{\mu}\partial/\partial y^{\mu}\ \gamma_{z\beta\lambda}{}^{\nu}\partial/\partial z^{\nu} + M]\psi_{2TT}{}^{\kappa\lambda} \qquad (8.1)$$

where $\gamma_y{}^{\mu}$ and $\gamma_z{}^{\mu}$ are Dirac matrices for the y and z coordinates respectively, and y and z are coordinates in r dimension space-times, M is the mass, and

$$\overline{\psi}_{kTT\alpha\beta} = -\psi_{kTT}{}^{\kappa\lambda\dagger}(\gamma_y{}^{0}\gamma_y{}^{5})_{\kappa\alpha}\,(\gamma_z{}^{0}\gamma_z{}^{5})_{\lambda\beta} \qquad (8.2)$$

following eq. 3.35, for i, k = 1, 2 where the subscripts y and y indicate Dirac spinors associated with the y and z coordinates.

The form of the complex conjugate, eq. 8.2, implies that the alternate possible expression (a summation) for $\mathcal{L}$, namely $\gamma_{y\alpha\kappa}{}^{\mu}\partial/\partial y^{\mu} + \gamma_{z\beta\lambda}{}^{\nu}\partial/\partial z^{\nu}$ is not appropriate.

The equations of motion are

$$[M^{-1}\gamma_y{}^{\mu}\partial/\partial y^{\mu}\ \gamma_z{}^{\nu}\partial/\partial z^{\nu} - M]\psi_{1TT} = 0 \qquad (8.2a)$$
$$[M^{-1}\gamma_y{}^{\mu}\cdot\partial/\partial y^{\mu}\ \gamma_z{}^{\nu}\partial/\partial z^{\nu} - M]\psi_{2TT} = 0 \qquad (8.2b)$$

We define subsidiary equations of motion

$$[\gamma_y{}^{\mu}\partial/\partial y^{\mu} - M]\psi_{jTT} = 0 \qquad (8.2c)$$
$$[\gamma_z{}^{\nu}\partial/\partial z^{\nu} - M]\psi_{jTT} = 0 \qquad (8.2d)$$

for j = 1, 2. Eqs. 8.2c and 8.2d imply eqs. 8.2a and 8.2b. Note eq. 8.2d is tachyonic by design.

One conjugate momentum is (with two spinor indices α, β)

$$\pi_{y1TT\alpha\beta} = \partial\mathcal{L}/\partial(\partial\psi_{1TT\alpha\beta}/\partial y^{0}) = -M^{-1}(\partial/\partial z^{\nu}\ \psi_{2TT}{}^{\dagger\kappa\lambda}\gamma_y{}^{5}{}_{\kappa\alpha}\,(\gamma_z{}^{0}\gamma_z{}^{5}\gamma_z{}^{\nu})_{\lambda\beta}$$

$$= -\psi_{2TT}^{\dagger\kappa\lambda}\gamma_y^5{}_{\kappa\alpha}\,(\gamma_z^0\gamma_z^5)_{\lambda\beta} \qquad (8.3)$$

after partial integrations (with surface terms set to zero) of

$$L = \int d^r y \int d^r z \; \mathscr{L}$$

using the subsidiary tachyonic equation of motion:

$$\gamma_z^\nu \partial/\partial z^\nu \, \psi_{2TT}^\dagger = M\psi_{2TT}^\dagger \qquad (8.4)$$

Similarly the other conjugate momenta may be found in a manner analogous to eqs. 5.3 – 5.7.

8.2 PseudoFermion Quantum Fields with a Tachyon Interior

A *tentative* free PseudoFermion TT wave function has the form:

$$\psi_{iTT\alpha\beta}(y, z) = \Sigma\Sigma \int dp^{r-1}\int dq^{r-1}\, N(p, q)\,[b_{iTT}(p, q,s_1,s_2)u_\alpha(p,s_1)u_{\beta TT}(q, s_2)\exp(-ip\cdot y - iq\cdot z) +$$

$$+ d_{iTT}^\dagger(p, q,s1,s2)v_\alpha(p,s1)v_{\beta TT}(q, s2)\exp(ip\cdot y + iq\cdot z)] \qquad (8.5)$$

plus Hermitean conjugates for i = 1, 2 where $N(p, q)$ is a normalization factor. Note both parts (y and z) of ψ_i are on the mass shell: $p^2 = M^2$ and $q^2 = -M^2$.

At this point we might attempt to complete the canonical quantization procedure in the conventional manner by Fourier expanding the quantum field and specifying anti-commutation relations for the Fourier component amplitudes. However the incompleteness of the set of plane waves, which are limited by the restrictions $|q| \geq M$ and $|p| \geq M$, causes the anti-commutator of the fields not to yield a $\delta^3(x - x')$. Thus the conventional approach fails to yield the required anti-commutation relations.

Other approaches: 1) decompose the tachyon fields into left-handed and right-handed parts and then second quantize each part; and 2) second quantize in light-front coordinates ($x^\pm = (x^0 \pm x^3)/\sqrt{2}$). These approaches also both fail.[69]

The only approach that does succeed[70] is to decompose the tachyon parts of the field into left-handed and right-handed parts and then second quantize the y and z coordinate part in light-front coordinates.

The quantized development of this case proceeds in a manner similar to that of chapter 5.

---

[69] See the first edition Blaha (2006) where these possibilities were considered and found to fail.
[70] Blaha (2006) discusses this case in detail.

9. Quadplex Fermion Bradyon-Tachyon SU(4) Strong Interaction Theory

Chapter 6 developed a Duplex Bradyon-Tachyon (BT) ElectroWeak Theory based on PseudoFermion fields where one (exterior) part corresponded to the Lab with y coordinates and the other (interior) part corresponded to the particle interior with z coordinates. The exterior field was always bradyonic; the interior field was bradyonic for charged leptons and up-type quarks, and tachyonic for neutral leptons and down-type quarks.

We now introduce a new *Quadplex Fermions* formulation for the Strong Interactions starting with an initial SU(4) symmetry group (later broken to SU(3)⊗U(1)) composed of three quarks and a lepton. We will again use the distinction between bradyons and tachyons to create a set of four Quadplex fields. Each Quadplex fermion field will consist of four parts where each part is either bradyonic or tachyonic. We associate each quadplex field with a fermion in a fundamental SU(4) representation of four dimensions.

We use b to signify a bradyonic field and t to signify a tachyonic field. We will use the up-type and down-type fermion sequences found in our previous books for quadplex fields:

<u>Up-Type Sequence</u>

$$e \leftrightarrow \psi_e = \psi_{1\uparrow} \sim b_y b_z t_u t_v \qquad (9.1)$$
$$u \leftrightarrow \psi_u = \psi_{2\uparrow} \sim b_y t_z t_u b_v$$
$$c \leftrightarrow \psi_c = \psi_{3\uparrow} \sim b_y b_z b_u t_v$$
$$t \leftrightarrow \psi_t = \psi_{4\uparrow} \sim b_y b_z b_u b_v$$

<u>Down-Type Sequence</u>

$$\nu_e \leftrightarrow \psi_e = \psi_{1\downarrow} \sim t_y t_z t_u t_v$$
$$\nu' \leftrightarrow \psi_{\nu'} = \psi_{1\downarrow} \sim b_y t_z t_u t_v$$
$$d \leftrightarrow \psi_d = \psi_{2\downarrow} \sim b_y t_z t_u b_v$$
$$s \leftrightarrow \psi_s = \psi_{3\downarrow} \sim b_y t_z b_u t_v$$
$$b \leftrightarrow \psi_b = \psi_{4\downarrow} \sim b_y t_z b_u b_v$$

where we use particle symbols to represent wave functions with coordinates indicated with subscripts. The ElectroWeak part of each fermion uses y and z coordinates of the type described in chapter 6, and the SU(4) part uses u and v coordinates (together with y and z parts). All four coordinate systems have four dimensions. The top-type fermion assignment in eq. 9.1 is *generally based on the ordering of quark masses with heavier fermions having more Bradyon parts and lighter fermions having more Tachyon parts.* A similar assignment may be made for down-type fermions. The list is modified to match ElectroWeak pairs with up-type bradyon matched with down-type tachyon except for the e – v (neutrino) case where the match is e - v'.

The z, u and v coordinate parts may be bradyonic or tachyonic since these parts are not directly relevant for the Physical LAB frame. The z, u and v coordinates are <u>not</u> "LAB" coordinates. They are internal to the particles.

The y part coordinates are the LAB coordinates. The z, u and v parts are not connected[71] to the LAB coordinates and do not support faster than light motion in the LAB frame.

The fermion wave functions generally differ. They may have bradyon or tachyon internal parts. As a result the SU(4) generators which are usually represented numerically must be generalized to have bradyon – tachyon projection operator factors. See eqs. 9.11 – 9.16 for their new form. The approach here, based on bradyon-Tachyon internal differences, is important for differentiating between fermions. It offers a fermion internal mechanism for understanding SU(4) (SU(3)⊗U(1)) interaction experimental data.

SL(2, C)⊗SU(2)⊗U(1)⊗S(4) Parts

9.1 SL(2, C)⊗SU(2)⊗U(1)⊗SU(4) Parts

The SL(2, C)⊗SU(2)⊗U(1) part of the UST has 8 real dimensions in the UST dimension array. The SU(4) part of the UST also has 8 real dimensions in the UST dimension array. See Fig. 2.1.

The SU(4) structure[72] may be embedded within a complex four fermion particle representations according to eq. 9.1. The SU(4) fundamental representation being complex has a part for its four real dimensions (u coordinates) and a part for the four imaginary dimensions (v coordinates). Both parts support bradyon and tachyon sectors. We denote the coordinate system for the real part as u. We denote the coordinate system for the imaginary part as v. The fermion sector combines the SU(4)-specific u and v coordinate systems with the y and z coordinates for the SL(2, C)[73]⊗SU(2) ElectroWeak sector described in chapter 6.

Thus the sets of coordinate systems are:[74]

<u>SL(2, C)⊗SU(2)</u>[75]		<u>SU(4)</u>		(9.2)
y	z	u	v	
Lab frame	Imaginary Part	Real Part	Imaginary Part	
Bradyon Always	Bradyon or Tachyon	Bradyon or Tachyon	Bradyon or Tachyon	

---

[71] Later we will discuss connecting the various parts together with interactions.
[72] The SU(4) group is broken to SU(3)⊗U(1).
[73] Signifies a Lorentz group SO⁺(1,3) representation.
[74] One might establish an alternate four dimension quaternion form y +iz + ju + kv of Lagrangian theory perhaps using the approaches of S. Adler and S. Frautschi.
[75] Signifies a Lorentz group SO⁺(1,3) representation.

Free PseudoQuantum Lagrangian for All Parts

A free *PseudoQuantum* Lagrangian for the combined symmetries is

$$\mathscr{L} = \overline{\psi}_{\uparrow 2p}{}^{\alpha\beta\rho\sigma}\mathscr{K}_{\uparrow pq}[M_{\uparrow}{}^{-3}\gamma_{y\alpha\kappa}{}^{\mu}\partial/\partial y^{\mu}\ \gamma_{z\beta\lambda}{}^{\nu}\partial/\partial z^{\nu}\ \gamma_{u\rho\tau}{}^{\mu}\partial/\partial u^{\mu}\ \gamma_{v\sigma\chi}{}^{\nu}\partial/\partial v^{\nu} - M_{\uparrow}]\psi_{\uparrow 1q}{}^{\kappa\lambda\tau\chi} +$$
$$+\ \mathscr{K}_{\uparrow pq}\overline{\psi}_{\uparrow 1p}{}^{\alpha\beta\rho\sigma}[-M_{\uparrow}{}^{-3}\gamma_{y\alpha\kappa}{}^{\mu}\partial/\partial y^{\mu}\ \gamma_{z\beta\lambda}{}^{\nu}\partial/\partial z^{\nu}\gamma_{u\rho\tau}{}^{\mu}\partial/\partial u^{\mu}\ \gamma_{v\sigma\chi}{}^{\nu}\partial/\partial v^{\nu} - M_{\uparrow}]\psi_{\uparrow 2q}{}^{\kappa\lambda\tau\chi} +$$
$$+\ \overline{\psi}_{\downarrow 2p}{}^{\alpha\beta\rho\sigma}\mathscr{K}_{\downarrow pq}[M_{\downarrow}{}^{-3}\gamma_{y\alpha\kappa}{}^{\mu}\partial/\partial y^{\mu}\ \gamma_{z\beta\lambda}{}^{\nu}\partial/\partial z^{\nu}\ \gamma_{u\rho\tau}{}^{\mu}\partial/\partial u^{\mu}\ \gamma_{v\sigma\chi}{}^{\nu}\partial/\partial v^{\nu} - M_{\downarrow}]\psi_{\downarrow 1q}{}^{\kappa\lambda\tau\chi} +$$
$$+\ \mathscr{K}_{\downarrow pq}\overline{\psi}_{\downarrow 1p}{}^{\alpha\beta\rho\sigma}[-M_{\downarrow}{}^{-3}\gamma_{y\alpha\kappa}{}^{\mu}\partial/\partial y^{\mu}\ \gamma_{z\beta\lambda}{}^{\nu}\partial/\partial z^{\nu}\ \gamma_{u\rho\tau}{}^{\mu}\partial/\partial u^{\mu}\ \gamma_{v\sigma\chi}{}^{\nu}\partial/\partial v^{\nu} - M_{\downarrow}]\psi_{\downarrow 2q}{}^{\kappa\lambda\tau\chi}$$

$$(9.3)$$

where ψ_1 and ψ_2 are PseudoQuantum field column vectors containing the various fields in combinations specified in eq. 9.5 below, where p and q designate fermions according to eq. 9.5 below, and where $\gamma_y{}^{\mu}$, $\gamma_z{}^{\mu}$, $\gamma_u{}^{\mu}$ and $\gamma_v{}^{\mu}$ are Dirac matrices for the y, z, u and v coordinates respectively. $M_{\uparrow}$ and $M_{\downarrow}$ are the 4×4 mass diagonal matrices of the up-type and down-type fermion sequences respectively.

The diagonal matrices $\mathscr{K}_{\uparrow}$ and $\mathscr{K}_{\downarrow}$ handle the numeric factors for the b and t content of the respective individual fermion particles in terms of their bradyon and tachyon content is for generic fermions:

$$\mathscr{K}_{\uparrow} = \begin{bmatrix} (-i)^2 & 0 & 0 & 0 \\ 0 & (-i)^2 & 0 & 0 \\ 0 & 0 & (-i)^3 & 0 \\ 0 & 0 & 0 & (-i)^4 \end{bmatrix} = \begin{bmatrix} -1 & 0 & 0 & 0 \\ 0 & -1 & 0 & 0 \\ 0 & 0 & i & 0 \\ 0 & 0 & 0 & 1 \end{bmatrix} \qquad (9.4)$$

$$\mathscr{K}_{\downarrow} = \begin{bmatrix} 1 & 0 & 0 & 0 \\ 0 & (-i)^2 & 0 & 0 \\ 0 & 0 & (-i)^2 & 0 \\ 0 & 0 & 0 & (-i)^3 \end{bmatrix} = \begin{bmatrix} 1 & 0 & 0 & 0 \\ 0 & -1 & 0 & 0 \\ 0 & 0 & -1 & 0 \\ 0 & 0 & 0 & i \end{bmatrix} \qquad (9.5)$$

The y and z ElectroWeak Sector Parts

Focusing on the SU(4) related parts, we begin by "factoring" out the y and z dependent parts using the subsidiary equations of motion similar to eq. 8.2c and 8.2d which have the form:[76]

$$[i\gamma_y{}^{\mu}\partial/\partial y^{\mu} - M]\psi = 0 \qquad\qquad (9.6)$$
$$[i\gamma_z{}^{\mu}\partial/\partial z^{\mu} - M]\psi = 0 \qquad\text{Bradyonic z field part}$$

or

$$[\gamma_z{}^{\mu}\partial/\partial z^{\mu} - M]\psi = 0 \qquad\text{Tachyonic z field part}$$

---

[76] We use conventional Quantum Field theory from this point in this chapter for convenience. It may be directly generalized to PseudoQuantum Field Theory.

where M represents diagonal mass matrices $M_\uparrow$ or $M_\downarrow$ for the up-type and down-type fermion sequences as the case may be here and in the following sections.

The u and v SU(4) Related Parts for Up-Type Fermions

The u and v parts of the four up-type fermion fields satisfy the subsidiary equations:

$$
\begin{aligned}
e: &\quad (T_u T_v)\psi_{T_u T_v} \equiv [-M^{-1}\gamma_u{}^\mu \partial/\partial u^\mu \; \gamma_v{}^\nu \partial/\partial v^\nu - M]\psi_{T_u T_v} = 0 \\
u: &\quad (T_u B_v)\psi_{T_u B_v} \equiv [iM^{-1}\gamma_u{}^\mu \partial/\partial u^\mu \; \gamma_v{}^\nu \partial/\partial v^\nu - M]\psi_{T_u B_v} = 0 \\
c: &\quad (B_u T_v)\psi_{B_u T_v} \equiv [iM^{-1}\gamma_u{}^\mu \partial/\partial u^\mu \; \gamma_v{}^\nu \partial/\partial v^\nu - M]\psi_{B_u T_v} = 0 \quad (9.7) \\
t: &\quad (B_u B_v)\psi_{B_u B_v} \equiv [-M^{-1}\gamma_u{}^\mu \partial/\partial u^\mu \; \gamma_v{}^\nu \partial/\partial v^\nu - M]\psi_{B_u B_v} = 0
\end{aligned}
$$

interpreting T and B similarly to the operators defined in chapter 6 in equations:

$$
B_z\psi = [i\gamma_z{}^\mu \partial/\partial z^\mu - M]\psi = 0 \tag{4.2c}
$$
$$
T_z\psi_T = [\gamma_z{}^\mu \partial/\partial z^\mu - M]\psi_T = 0 \tag{5.2c}
$$

Green's Functions

The four factors of M for each sequence appearing in eq. 9.3, 9.6 and 9.7 may be treated more generally as four differing mass matrices M_1, M_2, M_3, M_4 in the various equations such as eq. 9.7. The result would be different free field propagators for each of the four parts of a wave function.

The Green's function for the up-type part of eq. 9.3 is proportional to

$$
M_1 M_2 M_3 M_4 [\mathcal{K}_\uparrow^2 p\cdot\gamma_y \; q\cdot\gamma_z \; r\cdot\gamma_u \; s\cdot\gamma_v - M_1 M_2 M_3 M_4]^{-1}
$$

The Green's function for eq. 9.7 is proportional to

$$
\{[-ip\cdot\gamma_y - M_1][p\cdot\gamma_z - M_2]\,[r\cdot\gamma_u - M_3][-is\cdot\gamma_v - M_4]\}^{-1}
$$

Field Equations

The quadplex subsidiary up-type *free* field equations for eq. 9.7 may be expressed using

$$
đ_\uparrow(u, v) = \begin{bmatrix}
(T_u T_v) & 0 & 0 & 0 \\
0 & (T_u B_v) & 0 & 0 \\
0 & 0 & (B_u T_v) & 0 \\
0 & 0 & 0 & (B_u B_v)
\end{bmatrix} \tag{9.8}
$$

The operator $đ_\uparrow(u, v)$ is a 16×16 matrix operator with the 4×4 operator matrix parts shown above in eq. 9.8. Then the extended up-type free field equation is

$$
đ_\uparrow(u, v)\Psi_\uparrow(u, v) = 0 \tag{9.10}
$$

Down-type fermions may be similarly treated.

Minimal Tachyonic Transformations

9.2 Minimal Bradyon-Tachyonic Transformations

The most minimal tachyonic transformation maps wave functions between bradyon and tachyon forms. Following the conclusions of section 6.4 we use the transformation:

$$S_L(\Lambda_L(\omega, \mathbf{u})) = \exp(-i\omega_L\sigma_{0i}v_i/(2|\mathbf{v}|)) = \exp(-\omega_L\gamma^0\boldsymbol{\gamma}\cdot\mathbf{v}/(2|\mathbf{v}|))$$
$$= i\,\sinh(\omega/2)I + i\,\cosh(\omega/2)\gamma^0\boldsymbol{\gamma}\cdot\mathbf{u} \qquad (3.26)$$

which becomes the minimal transformation

$$S_L(\Lambda_L(\omega, \mathbf{u})) = i\gamma^0\boldsymbol{\gamma}\cdot\mathbf{u} = i\gamma^0\gamma_3 \qquad (6.32)$$
$$S = i\gamma^0\gamma_3$$
$$S_u = i\gamma_u{}^0\gamma_{u3}$$
$$S_v = i\gamma_v{}^0\gamma_{v3}$$

when $\omega = 0$ and $u_z = \gamma_3$ and the inverse transformation

$$S_L^{-1}(\Lambda_L(\omega, \mathbf{u})) = \gamma^2\gamma^0\kappa^{-1}S_L{}^\dagger\kappa\gamma^0\gamma^2 = \gamma^2\gamma^0 S_L{}^\tau\gamma^0\gamma^2 = \exp(\omega_L\gamma^0\boldsymbol{\gamma}\cdot\mathbf{v}/(2|\mathbf{v}|))$$
$$= i\,\sinh(\omega/2)I - i\,\cosh(\omega/2)\gamma^0\boldsymbol{\gamma}\cdot\mathbf{u} \qquad (3.27)$$

which becomes the minimal transformation

$$S_L^{-1}(\Lambda_L(\omega, \mathbf{u})) = -i\,\gamma^0\boldsymbol{\gamma}\cdot\mathbf{u} = -i\gamma^0\gamma_3 \qquad (6.33)$$
$$S^{-1} = -i\gamma^0\gamma_3$$
$$S_u^{-1} = -i\gamma_u{}^0\gamma_{u3}$$
$$S_v^{-1} = -i\gamma_v{}^0\gamma_{v3}$$

when $\omega = 0$ and $u_z = \gamma_3$. Note:the squares of the minimal transformations are:

$$S_L(\Lambda_L(0, \mathbf{u}))^2 = S^2 = -I \qquad (6.33a)$$
$$S_L^{-1}(\Lambda_L(0, \mathbf{u}))^2 = (S^{-1})^2 = -I \qquad (6.33b)$$

9.3 SU(4) Generators

The generators of the SU(4) Lie algebra are 15 matrices in its four dimension fundamental representation. These matrices are traceless, Hermitian matrices. We will denote them as T_1, T_2, ... , T_{15}. For example we can represent T_1 with:

$$T_1 = \begin{bmatrix} 0 & 1 & 0 & 0 \\ 1 & 0 & 0 & 0 \\ 0 & 0 & 0 & 0 \\ 0 & 0 & 0 & 0 \end{bmatrix} \tag{9.11}$$

In the SU(4) fundamental representation. We will represent the kth element of these matrices with

$$T_{kij} \tag{9.12}$$

where i labels columns and j labels rows.

Normally Strong interaction terms had the form

$$\overline{\psi}_i(y, z, u, v)A_kT_{kij}\psi_j(y, z, u, v) \tag{9.13}$$

summed over j and k where $\psi_j(u, v)$ is the jth row in eq. 9.9. We now wish to take account of the difference between the wave functions in the four rows by modifying T_{kij} using the above minimal bradyon-tachyon transformations to form new matrices S_{kij} specified by eq. 9.15 below. For example we replace T_1 with

$$S_1 = \begin{bmatrix} 0 & i\gamma_v{}^0\gamma_{v3} & 0 & 0 \\ -i\gamma_v{}^0\gamma_{v3} & 0 & 0 & 0 \\ 0 & 0 & 0 & 0 \\ 0 & 0 & 0 & 0 \end{bmatrix} \tag{9.14}$$

The components are multiplied by S and S^{-1} respectively.

In general we may define

For $i \neq j$
$$S_{kij} = T_{kij}T^S{}_{ij} \tag{9.15}$$

For $i = j$
$$S_{kij} = 0$$

where the $T^S{}_{ij}$ elements are given by

$$T^S = \begin{bmatrix} 0 & S_v & S_u & S_uS_v \\ S_v{}^{-1} & 0 & S_uS_v{}^{-1} & S_u \\ S_u{}^{-1} & S_u{}^{-1}S_v & 0 & S_v \\ S_u{}^{-1}S_v{}^{-1} & S_u{}^{-1} & S_v{}^{-1} & 0 \end{bmatrix} \tag{9.16}$$

Note the Hermiticity and traceless conditions of S_k for all k.

Now a quadplex Strong interaction term has the form

$$\overline{\psi}_i(y, z, u, v)A_kS_{kij}\psi_j(y, z, u, v) \tag{9.17}$$

The results of this section may be directly applied to the four down-type fermions.

These results may be combined with the Subluminal-Superluminal ElectroWeak results of chapter 6 to obtain a quadplex Subluminal-Superluminal Standard Model.

9.4 ElectroWeak Interaction of SU(4) Fundamental Fermions

The fundamental fermions interact with the charged ElectroWeak vector bosons $A^\pm$ (and the neutral vector boson A^0). The forms of some of the interaction terms for the quark parts of each sequence are

$$\overline{\psi_{\uparrow 2}}\, W^+ \, \psi_{\downarrow 2} + \overline{\psi_{\downarrow 2}}\, W^- \, \psi_{\uparrow 2}$$

where we use the PseudoQuantum field 2 for interaction terms as we have since the 1970's. The lepton part of each sequence interacts in a "retrograde" fashion:

$$\overline{\psi_{\uparrow 2e}}\, W_{ev'}{}^- \, \psi_{\downarrow 2v'} + \overline{\psi_{\downarrow 2v'}}\, A W_{'e}{}^+ \, \psi_{\uparrow 2e}$$

due to the negative charge of the electron. There are Z and electromagnetic interactions as well. The W^+, W^- and W^0 terms have subluminal-superluminal factors as in:

$$D^\mu(y) = \partial^\mu + ig\mathbf{t}(S_{zL})W^\mu(y) + ig't_0(S_{zL})W_0{}^\mu(y) \qquad (6.37)$$
$$D^i(y) = \partial^i + ig\mathbf{t}(S_{zL})W^i(y) \qquad (6.37a)$$
$$D^0(y) = \partial^0 + ig't_0(S_{zL})W_0{}^0(y) \qquad (6.37b)$$

Thus we have introduced the bradyon-tachyon formalism in the ElectroWeak and Strong Interaction formulation to enable a theoretical framework for the appearance of tachyonic behavior within fermions without the usual diseases associated with tachyons.

9.5 Eight Dimension Form of the ElectroWeak and Strong Interactions

The ElectroWeak and the Strong Interactions can be put in an eight dimension matrix form. We define an 8-vector representing the up-type and down-type sequences of fermions:

$$\Psi = \begin{bmatrix} e \\ u \\ c \\ t \\ v' \\ d \\ s \\ b \end{bmatrix} \qquad (9.18)$$

We next define the Strong Interaction SU(4) Lagrangian term:

$$\mathcal{L}_{\text{Strong}} = \overline{\Psi} \begin{bmatrix} \begin{array}{c|c} \begin{matrix} \text{SU(4)} \\ \text{Representation} \end{matrix} & 0 \\ \hline 0 & \begin{matrix} \text{SU(4)} \\ \text{Representation} \end{matrix} \end{array} \end{bmatrix} \Psi \tag{9.19}$$

We next define the ElectroWeak SU(2)⊗U(1) Lagrangian term:

$$\mathcal{L}_{\text{ElectroWeak}} = \overline{\Psi} \begin{bmatrix} \begin{array}{c|c} \begin{matrix} \text{SU(2)⊗U(1)} \\ \text{Neutral} \\ \text{Vector Fields} \end{matrix} & \begin{matrix} \overset{A^-\ 0\ 0\ 0}{} \\ \text{SU(2)⊗U(1)} \\ A^+ \text{ Charged} \\ \text{Vector Fields} \end{matrix} \\ \hline \begin{matrix} \overset{A^+\ 0\ 0\ 0}{} \\ \text{SU(2)⊗U(1)} \\ A^- \text{ Charged} \\ \text{Vector Fields} \end{matrix} & \begin{matrix} \text{SU(2)⊗U(1)} \\ \text{Neutral} \\ \text{Vector Fields} \end{matrix} \end{array} \end{bmatrix} \Psi \tag{9.20}$$

Note that the electron and neutrino components have charged vector boson factors opposite to the other charged vector boson factors due to their charge (or lack of charge).

We now define the diagonal Lagrangian fermion mass term:

$$\mathcal{L}_{\text{Mass}} = \overline{\Psi} \begin{bmatrix} e & & & & & & & \\ & u & & & & & & \\ & & c & & & & & \\ & & & t & & & & \\ & & & & \nu' & & & \\ & & & & & d & & \\ & & & & & & s & \\ & & & & & & & b \end{bmatrix} \Psi \tag{9.21}$$

where the symbols indicate mass values and with off-diagonal elements being zeroes. The values of the masses and their representation as products of terms containing π, e and 2 are[77]

	SEQUENCE 1					**SEQUENCE 2**			
	e	u	c	t		ν'	d	s	b
Mass:	$0.511{\times}10^{-3}$	$1.80{\times}10^{-3}$	1.28	171		8.4×10^{-5}	$4.24{\times}10^{-3}$	$102{\times}10^{-3}$	4.34
Multiplier:		$2^5\pi$	$2^5\pi$	$2^5\pi$		32	32		32

$$1 \qquad \nu_e = \alpha_0^2/e^2 \;\; \nu' = e^3 2^{-39} \text{ GeV}/c^2$$
$$1 \qquad \nu' = e^2(\alpha_0^2)^{-1}\,\nu_e \cong e\,2^{-15}\text{ GeV}/c^2$$
$$1 \qquad e = r_1\,\nu' = \pi e 2^{-14}\text{ GeV}/c^2$$
$$1 \qquad d = r_2 e = r_2 r_1 \nu' = \pi e 2^{-11}\text{ GeV}/c^2$$
$$\tfrac{3}{4} \qquad u = r_3 d = r_3 r_2 r_1 \nu' = \tfrac{3}{4}\times\pi e 2^{-12}\text{ GeV}/c^2$$
$$\tfrac{3}{4} \qquad c = r_4 u = r_4 r_3 r_2 r_1 \nu' = \tfrac{3}{4}\times\pi^2 e 2^{-4}\text{ GeV}/c^2$$
$$\tfrac{3}{4} \qquad s = r_5 c = r_5 r_4 r_3 r_2 r_1 \nu' = \tfrac{3}{4}\times\pi e 2^{-6}\text{ GeV}/c^2$$
$$\tfrac{1}{2} \qquad t = r_6 s = r_6 r_5 r_4 r_3 r_2 r_1 \nu' = 2\pi^3 e\text{ GeV}/c^2$$
$$\tfrac{1}{2} \qquad b = r_7 t = r_7 r_6 r_5 r_4 r_3 r_2 r_1 \nu' = \pi e 2^{-1}\text{ GeV}/c^2$$

Figure 2.8. Revised form of fermion masses using $\nu' \cong e2^{-15}$ GeV/c^2. From Blaha (2024i).

It is evident from eq. 9.20 and the sequences of fermion masses that ElectroWeak $SU(2){\otimes}U(1)$ is responsible for the breakdown of $SU(4)$ to $SU(3){\otimes}U(1)$.

9.6 Consequences of Subluminal-Superluminal Strong Interaction SU(4)

Chapter 9 gives a new formulation of The Quadplex Subluminal-Superluminal Standard Model within the framework of UST and Cosmos Theory. This theory embodies the $SO(1, 3)^+{\otimes}SU(2){\otimes}U(1){\otimes}S(4)$ part of the first layer of the r = 4 dimension array of Cosmos Theory. See Blaha (2024e) for details.

This formulation has several advantages:

1. It explains the apparent faster than light behavior implied by certain fermion spin phenomena in fermion particle models. See section 6.7.

2. It opens the possibility of superluminal Physics (and possibly travel) through bradyon-tachyon interactions generating "pure" tachyon (and bradyon) motion. See chapter 12.

---

[77] From Blaha (2024i).

10. Higgs-less Bradyon-Tachyon PseudoBoson Quantum Theory

We now turn to develop a *duplex* Bradyon-Tachyon Higgs-less PseudoBoson Quantum Theory with a view towards a perturbation theory.[78] We begin with free PseudoBoson theory. Then vector gauge boson theories follow.

10.1 Free Higgs-less Bradyon-Tachyon PseudoBoson Formalism

We define a free real-valued scalar Higgs-less (massless) duplex Bradyon-Tachyon PseudoBoson formalism in a manner analogous to the PseudoFermion formalism with

$$\varphi_i(y, z) \tag{10.1}$$

where y and z are independent coordinates with each in r space-time dimensions, and i =1, 2 labels the PseudoQuantum field.[79] The y coordinates are for bradyons; the z coordinates are for tachyons.

We define a free real-valued scalar duplex PseudoQuantum PseudoBoson Lagrangian[80] with two related wave functions φ_1 and φ_2 that are functions of two sets of coordinates in r space-time dimensions, y and z:

$$\mathcal{L} = -m^{-2}\partial^2\varphi_1/\partial y^\mu \partial z^\nu\, \partial^2\varphi_2/\partial y_\mu \partial z_\nu - m^2\varphi_1\varphi_2 \tag{10.2}$$

where m is the mass. Using partial integrations again we find the equations of motion:

$$-m^{-2}\Box_y\Box_z\, \varphi_i + m^2\varphi_i = 0 \tag{10.3}$$

---

[78] Chapter 6 of Blaha (2022c) discusses wave functions with multiple coordinates. Blaha (2023c), which is epitomized here, discusses PseudoBoson theory. This chapter extends those chapters to bradyon-tachyon wave functions.

[79] Appendix C of Blaha (2016f) presents a first quantized PseudoQuantum theory CQ Mechanics that embodies both classical and quantum theory. **That theory is the non-relativistic quantum mechanics limit of the PseudoFermion theory developed here.** CQ Mechanics has two sets of coordinates that combine to create a generalization of conventional Quantum Mechanics. It has applications in a generalized Feynman path integral formalism, a generalized Schrödinger equation, a generalized Boltzmann equation, the Fokker-Planck equation, a generalized approach to quantum and classical chaos, and to quantum entanglement as well as semi-quantum entanglement. Our "Pseudo" formalisms apply to both Quantum Field Theory and Quantum Mechanics as well as the path integrals, the Fokker-Planck equation and the Boltzmann equation. In these applications there is a clear almost continuous transition from the quantum to the classical sectors.

[80] The choice of lower order derivative terms leads to unsatisfactory results. The fourth order derivatives in eq. 8.3 raise the issue of a linear potential. See S. Blaha, Phys. Rev. **D10**, 4268 (1974).

for $i = 1, 2$ where the y and z subscripts indicate derivatives with respect to y and z respectively, and where $\Box_z = \partial/\partial z_\mu \, \partial/\partial z^\mu$ and similarly for $\Box_y$. Assuming the z coordinates behavior of the fields is tachyonic the subsidiary equations of motion are

$$\Box_y \varphi_i + m^2 \varphi_i = 0 \qquad (10.4)$$
$$\Box_z \varphi_i - m^2 \varphi_i = 0 \qquad (10.5)$$

for $i = 1, 2$ in an r dimension space-time. They imply eq. 10.3 above.

One conjugate momentum is

$$\pi_{y1} = \partial \mathcal{L}/\partial(\partial\varphi_1/\partial y^0) = m^{-2}\, \partial/\partial z^\mu \, \partial^2\varphi_2/\partial y^0 \partial z_\mu = m^{-2}\Box_z\, \partial\varphi_2/\partial y^0 \qquad (10.6)$$
$$= -\partial\varphi_2/\partial y^0$$

by eq. 10.5 after partial integrations (with surface terms having the value zero) where

$$L = \int d^r y \int d^r z \, \mathcal{L} \qquad (10.7)$$

Similarly the other conjugate momenta are

$$\pi_{z1} = \partial \mathcal{L}/\partial(\partial\varphi_1/\partial z^0) = \partial\varphi_2/\partial z^0 \qquad (10.8)$$
$$\pi_{y2} = \partial \mathcal{L}/\partial(\partial\varphi_2/\partial y^0) = -\partial\varphi_1/\partial y^0$$
$$\pi_{z2} = \partial \mathcal{L}/\partial(\partial\varphi_2/\partial z^0) = \partial\varphi_1/\partial z^0$$

The form of the conjugate momenta[81] implies the selection of the non-zero equal time commutators:

$$[\partial\pi_{zi}(\mathbf{y}, y^0, \mathbf{z}, z^0)/\partial y^0, \ \varphi_j(\mathbf{y'}, y^0, \mathbf{z'}, z^0)] = (1 - \delta_{ij})\delta^{r-1}(\mathbf{y} - \mathbf{y'})\delta^{r-1}(\mathbf{z} - \mathbf{z'}) \qquad (10.9)$$
$$[\partial\pi_{yi}/(\mathbf{y}, y^0, \mathbf{z}, z^0)\partial z^0, \ \varphi_j(\mathbf{y'}, y^0, \mathbf{z'}, z^0)] = (1 - \delta_{ij})\delta^{r-1}(\mathbf{y} - \mathbf{y'})\delta^{r-1}(\mathbf{z} - \mathbf{z'})$$

in r space-time dimensions for $i, j = 1, 2$.

The free duplex PseudoFermion wave function has the form:

$$\varphi_i(y, z) = \int dp^{r-1} \int dq^{r-1} \, N(p, q) \, [a_i(p, q) \exp(-ip \cdot y - iq \cdot z) + a_i^\dagger(p, q) \exp(ip \cdot y + iq \cdot z)] \qquad (10.10)$$

plus Hermitean conjugates for $i = 1, 2$ where $N(p, q)$ is a normalization factor.

The creation and annihilation operators satisfy the commutation relations:

$$[a_i(p, q), a_j^\dagger(p', q')] = (1 - \delta_{ij})\, \delta^{r-1}(\mathbf{p} - \mathbf{p'})\delta^{r-1}(\mathbf{q} - \mathbf{q'}) \qquad (10.11)$$

---

[81] Taking account of $y - z$ symmetry.

The other commutators have zero values.

10.2 Free Duplex U(1) Vector PseudoBoson Formalism

This formalism is for electromagnetic-like duplex vector fields with a gauge transformation. The Bradyon-Tachyon distinction is irrelevant in this case since the fields are massless. The y coordinates are formally bradyonic and the z coordinates are formally tachyonic.

The duplex vector fields are $A_i^\mu(y, z)$ for $i = 1, 2$. The field strengths are

$$F^{1y}{}_{1\mu\nu\alpha} = \partial_{z\alpha} (\partial_{yv} A_{1\mu} - \partial_{y\mu} A_{1v})$$
$$F^{2y}{}_{1\mu\nu\alpha} = \partial_{z\alpha} (\partial_{yv} A_{2\mu} - \partial_{y\mu} A_{2v})$$

(10.12)

$$F^{1z}{}_{1\mu\nu\alpha} = \partial_{ya} (\partial_{zv} A_{1\mu} - \partial_{z\mu} A_{1v})$$
$$F^{2z}{}_{1\mu\nu\alpha} = \partial_{ya} (\partial_{zv} A_{2\mu} - \partial_{z\mu} A_{2v})$$

due to the need for gauge invariance. The additional derivatives introduced in eq. 10.12 are needed to preserve $y - z$ symmetry in the equal time commutators of eq. 10.20. Thus eq. 10.12 and the complete vector PseudoBoson Lagrangian have gauge invariance separately in the y and z coordinates:

<u>y coordinates</u>
$$A_{i\mu} \rightarrow A_{i\mu} + \partial_{y\mu}\Lambda(y)$$
<u>z coordinates</u>
$$A_{i\mu} \rightarrow A_{i\mu} + \partial_{z\mu}\Lambda(z)$$

for $i = 1, 2$.

The free duplex vector PseudoBoson Lagrangian is

$$\mathcal{L} = -\tfrac{1}{2}(F^{1y}{}_{\mu\nu\alpha} F^{2y\ \mu\nu\alpha} + F^{1z}{}_{\mu\nu\alpha} F^{2z\ \mu\nu\alpha})$$

(10.13)

Using partial integrations we find the equations of motion:

$$\Box_y^2 A_1^\mu = \Box_z^2 A_1^\mu = 0$$

(10.14)

$$\Box_y^2 A_2^\mu = \Box_z^2 A_2^\mu = 0$$

(10.15)

where $\Box_a^2 = (\partial_{a\mu}\partial_a^\mu)^2$ for $a = y, z$ in the radiation gauge where $A_1^0 = A_2^0 = 0$ and

$$\partial_{yi} A_2^i = \partial_{zi} A_2^i = 0$$

(10.16)

where the y and z subscripts indicate derivatives with respect to y and z respectively.

The conjugate momenta are

$$\pi_{y1}^\mu = \partial \mathcal{L} / \partial(\partial_y^0 A_{1\mu}) = \partial_{z\alpha} F^{2y\ \mu 0\alpha}$$

(10.17)

$$\pi_{z1}{}^{\mu} = \partial \mathcal{L}/\partial(\partial A_{1\mu}/\partial z^0) = \partial_{y\alpha} F^{2z\,\mu 0\alpha}$$

and

$$\pi_{y2}{}^{\mu} = \partial \mathcal{L}/\partial(\partial A_{2\mu}/\partial y^0) = \partial_{z\alpha} F^{1y\,\mu 0\alpha}$$

$$\pi_{z2}{}^{\mu} = \partial \mathcal{L}/\partial(\partial A_{2\mu}/\partial z^0) = \partial_{y\alpha} F^{1z\,\mu 0\alpha}$$

(10.18)

after partial integrations of

$$L = \int d^r y \int d^r z \; \mathcal{L}$$

(10.19)

with surface terms having the value zero.

The form of the conjugate momenta implies the only non-zero equal time commutators are:

$$[\partial\pi_{ya}{}^{\rho}(\mathbf{y}, y^0, \mathbf{z}, z^0)/\partial z^0, \; A_b{}^{\sigma}(\mathbf{y'}, y^0, \mathbf{z'}, z^0)] = (1 - \delta_{ab})\delta^{r-1\,\rho\sigma\,tr}(\mathbf{y} - \mathbf{y'})\delta^{r-1\,\rho\sigma\,tr}(\mathbf{z} - \mathbf{z'})$$ (10.20)

$$[\partial\pi_{za}{}^{\rho}(\mathbf{y}, y^0, \mathbf{z}, z^0)/\partial y^0, \; A_b{}^{\sigma}(\mathbf{y'}, y^0, \mathbf{z'}, z^0)] = (1 - \delta_{ab})\delta^{r-1\,\rho\sigma\,tr}(\mathbf{y} - \mathbf{y'})\delta^{r-1\,\rho\sigma\,tr}(\mathbf{z} - \mathbf{z'})$$

taking account of $y - z$ symmetry in r space-time dimensions for a, b = 1, 2 and ρ, σ = 1, 2, 3, ... , r – 1 where

$$\delta^{r-1\,\rho\sigma\,tr}(\mathbf{y} - \mathbf{y'}) = (2\pi)^{-r+1} \int d^{r-1}k \; e^{ik\cdot(\mathbf{y}-\mathbf{y'})} (\delta^{\rho\sigma} - k^\rho k^\sigma/k^2)$$

The free duplex PseudoBoson vector wave function has the form:

$$A_i{}^{\rho}(y, z) = \sum_{\lambda_1} \sum_{\lambda_2} \int dp^{r-1}\int dq^{r-1} \; N(p, q) \, [a_i(p, q, \lambda_1,\lambda_2)\varepsilon^{\rho}(p, q, \lambda_1, \lambda_2) \exp(-ip\cdot y - iq\cdot z) +$$

$$+ a_i{}^{\dagger}(p, q, \lambda_1,\lambda_2) \, \varepsilon^{\rho}(p, q, \lambda_1, \lambda_2) \exp(ip\cdot y + iq\cdot z)]$$

(10.21)

for i = 1, 2 where N(p, q) is a normalization factor. The unit vectors $\varepsilon(p, q, \lambda_1, \lambda_2)$ satisfy

$$\mathbf{p}\cdot\varepsilon = \mathbf{q}\cdot\varepsilon = 0$$

(10.22a)

$$\mathbf{p}\cdot\varepsilon = \mathbf{q}\cdot\varepsilon = 0$$

(10.22b)

$$\varepsilon(p, q, \lambda_1, \lambda_2)\cdot\varepsilon(p, q, \lambda_1', \lambda_2') = \delta_{\lambda_1'\lambda_1'}\delta_{\lambda_2'\lambda_2'}$$

(10.22c)

The creation and annihilation operators satisfy the commutation relations:

$$\{a_a(p, q, \lambda_1, \lambda_2), a_b{}^{\dagger}(p', q', \lambda_1', \lambda_2')\} = (1 - \delta_{ab})\delta_{\lambda_1,\lambda_1'}\delta_{\lambda_2,\lambda_2'}\delta^{r-1}(\mathbf{p} - \mathbf{p'})\delta^{r-1}(\mathbf{q} - \mathbf{q'})$$

(10.23)

The other commutators have zero values.

10.2.1 Quark Confining Linear Potential

The fourth order equations of motion (eqs. 10.14 and 10.15) lead to Green's functions have a k^{-4} momentum space behavior, which implies a confining linear potential if this model is used to consider quark confinement as we did in 1974.[82]

The k^{-4} propagator behavior emerges if we set $y = z$ in the Lagrangian of eq. 10.13 above:

$$\mathscr{L} = - F^{1y}{}_{\mu\nu\alpha} \, F^{2y\,\mu\nu\alpha} \tag{10.24}$$

which leads to

$$\Box_y^2 \, A_2{}^\mu = 0 \tag{10.25}$$

in the radiation gauge where $A_1{}^0 = A_2{}^0 = 0$

$$\partial_{y\rho} \, A_2{}^\rho = 0 \tag{10.26}$$

and

$$\Box_y^2 \, A_1{}^\mu = 0 \tag{10.27}$$

in the radiation gauge where

$$\partial_{y\rho} \, A_1{}^\rho = 0 \tag{10.28}$$

where $\Box_y^2 = (\partial/\partial y_\mu \, \partial/\partial y^\mu)^2$.

10.3 Non-Abelian Duplex Vector PseudoBoson Formalism

Non-Abelian PseudoQuantum gauge field formalisms[83] were developed by this author in the 1970s. These formalisms implemented quark confinement with a linear potential. The non-Abelian vector PseudoBoson theory would also support a quark confinement model with a linear confining potential.

This formalism for the non-Abelian gauge vector fields of a duplex vector gauge boson has the form $A_{i\mu}(y, z)$ for $i = 1, 2$. The Bradyon-Tachyon distinction is again irrelevant in the free fields case since the fields are massless. The y coordinates are formally bradyonic and the z coordinates are formally tachyonic.

The field strengths are

$$F^{1y}{}_{1\mu\nu\alpha} = \partial_{z\alpha} \left(\partial_{y\nu} A_{1\mu} - \partial_{y\mu} A_{1\nu} + gA_{1\mu} \times A_{1\nu} \right) \tag{10.29}$$
$$F^{2y}{}_{1\mu\nu\alpha} = \partial_{z\alpha} \left(\partial_{y\nu} A_{2\mu} - \partial_{y\mu} A_{2\nu} + gA_{1\mu} \times A_{2\nu} - gA_{1\nu} \times A_{2\mu} \right)$$
$$F^{1z}{}_{1\mu\nu\alpha} = \partial_{y\alpha} \left(\partial_{z\nu} A_{1\mu} - \partial_{z\mu} A_{1\nu} + gA_{1\mu} \times A_{1\nu} \right)$$
$$F^{2z}{}_{1\mu\nu\alpha} = \partial_{y\alpha} \left(\partial_{z\nu} A_{2\mu} - \partial_{z\mu} A_{2\nu} + gA_{1\mu} \times A_{2\nu} - gA_{1\nu} \times A_{2\mu} \right)$$

where $\times$ indicates a commutator. The free duplex vector PseudoBoson Lagrangian[84] is

$$\mathscr{L} = \tfrac{1}{2} \left(F^{1y}{}_{\mu\nu\alpha} \cdot F^{2y\,\mu\nu\alpha} - F^{1z}{}_{\mu\nu\alpha} \cdot F^{2z\,\mu\nu\alpha} \right) \tag{10.30}$$

where $\cdot$ indicates matrix multiplication.

---

[82] Explicit quark confinement in S. Blaha, Phys. Rev. **D10**, 4268 (1974).

[83] S. Blaha, Phys. Rev. **D10**, 4268 (1974); Phys. Rev. **D11**, 2921 (1975); Il Nuovo Cimento **49A**, 35 (1979).

[84] S. Blaha, Phys. Rev. **D11**, 2921 (1975) develops the formalism for this Lagrangian, which is eq. 17 in that paper, for the case where $z = y$. Eq. 98 of the paper has the quartic form $\Box\nabla^2 A_1{}^0 = g\lambda^2 J^0$, which indicates a k^{-4} gauge field propagator and thereby a linear confining potential.

The free part of $\mathcal{L}$ (truncated to quadratic terms in $A_{1\mu}$ and $A_{2\mu}$) leads to the wave function

$$A_{i\alpha}^{\rho}(y, z) = \sum_{\lambda_1} \sum_{\lambda_2} \int dp^{r-1} \int dq^{r-1}\, N(p, q)\, [a_{i\alpha}(p, q, \lambda_1, \lambda_2)\varepsilon^{\rho}(p, q, \lambda_1, \lambda_2)\exp(-ip \cdot y - iq \cdot z) +$$

$$+\, a_{i\alpha}^{\dagger}(p, q, \lambda_1, \lambda_2)\, \varepsilon^{\rho}(p, q, \lambda_1, \lambda_2)\exp(ip \cdot y + iq \cdot z)]   \quad (10.31)$$

using a development similar to the U(1) case above. The index α is the symmetry group index.

Eq. 10.30 leads to a k^{-4} gluon propagator and a linear quark confining potential if $z = y$.

10.4 PseudoGraviton Duplex PseudoQuantum Gravity

The duplex graviton fields in the weak field sector have the form $h_i^{\mu\nu}(y, z)$ for i = 1, 2. The Bradyon-Tachyon distinction is again irrelevant since the fields are massless. The y coordinates are formally bradyonic and the z coordinates are formally tachyonic.

We use a PseudoQuantum formulation by defining two metrics

$$g_{1\mu\nu} \cong \eta_{\mu\nu} + h_{1\mu\nu} \quad (10.32)$$
$$g_{2\mu\nu} \cong \eta_{\mu\nu} + h_{2\mu\nu} \quad (10.33)$$

where

$$|h_{1\mu\nu}| \ll 1 \quad (10.34)$$
$$|h_{2\mu\nu}| \ll 1$$

with

$$\partial_\mu h_1{}^\mu{}_\nu = \tfrac{1}{2}\partial_\nu h_1{}^\mu{}_\mu \quad (10.35)$$
$$\partial_\mu h_2{}^\mu{}_\nu = \tfrac{1}{2}\partial_\nu h_2{}^\mu{}_\mu \quad (10.36)$$

Neglecting quartic and other higher order terms we are led to the free dynamic equations

$$\Box\, h_i{}^{\mu\nu} = -16\pi G S^{\mu\nu} \quad (10.37)$$

for i = 1, 2 where $S^{\mu\nu}$ is the traceless energy-momentum tensor. The form of eq. 10.37 follows from the model duplex Lagrangian

$$\mathcal{L} = \partial h_1{}^{\mu\nu}/\partial y^\alpha\, \partial h_{2\mu\nu}/\partial y_\alpha + 16\pi G S^{\mu\nu}h_{2\mu\nu} + 16\pi G S^{\mu\nu}h_{1\mu\nu} \quad (10.38)$$

We now will introduce a second set of tachyonic coordinates z^μ with the view that our four dimension universe may perhaps have an additional set of coordinates giving the universe a complex set of dimensions.

We then require that the principle $y \leftrightarrow z$ enunciated in section 10.1 holds. The duplex Lagrangian then must have the form:

$$\mathscr{L} = M^{-2}\partial^2 h_1{}^{\mu\nu}/\partial y^\alpha\,\partial z^\beta\,\partial^2 h_{2\mu\nu}/\partial y_\alpha\,\partial z_\beta\ + 16\pi G S^{\mu\nu} h_{2\mu\nu} + 16\pi G S^{\mu\nu} h_{1\mu\nu} \qquad (10.39)$$

where $h_i{}^{\mu\nu} = h_i{}^{\mu\nu}(y, z)$. After partial integrations we find

$$M^{-2}\square_y\square_z\, h_i{}^{\mu\nu} = -16\pi G S^{\mu\nu} \qquad (10.40)$$

for $i = 1, 2$. If we let

$$G = g_0{}^2/M^2$$

then

$$\square_y\square_z\, h_i{}^{\mu\nu} = -16\pi\, g_0{}^2 S^{\mu\nu} \qquad (10.41)$$

The dimensional constant G is replaced by a dimensionless constant g_0 thus "explaining" the G coupling constant's mass^{-2} dependence.

In addition, if $y = z$ then a quartic derivative equation emerges that might have some relevance for the ongoing quandary over gravitation.

$$\square_y{}^2\, h_i{}^{\mu\nu} = -16\pi\, g_0{}^2 S^{\mu\nu} \qquad (10.42)$$

11. Higgsian Duplex Bradyon-Tachyon PseudoBoson Quantum Theory

In this chapter we consider massive Duplex Higgs Bradyon-Tachyon PseudoBoson Quantum Field Theories that embody both bradyon and tachyon parts. This combination has tachyon features combined with bradyon parts that give bosons a stability that prevents catastrophic condensation.

11.1 Free Duplex Higgsian Bradyon-Tachyon PseudoBoson Formalism

We define a free real-valued duplex scalar Higgsian (massive) Bradyon-Tachyon scalar PseudoBoson formalism in a manner analogous to the PseudoFermion formalism with

$$\varphi_i(y, z) \tag{11.1}$$

where y and z are independent coordinates with each in r space-time dimensions, and i =1, 2 labels the PseudoQuantum field.[85] The y coordinate are for bradyons; the z coordinates are for tachyons.

We define a free complex scalar Duplex PseudoQuantum PseudoBoson Lagrangian[86] with two related wave functions φ_1 and φ_2 that are functions of two sets of coordinates in r space-time dimensions, y and z. We will use the model duplex Lagrangian

$$\mathscr{L} = -\partial^2\varphi_1/\partial y^\mu \partial^2\varphi_2/\partial y_\mu \partial z_\nu \ - \ (\lambda\varphi_1^*\varphi_1 - \varphi_0^2)(\lambda\varphi_2^*\varphi_2 - \varphi_0^2) + \text{C.C.} \tag{11.2}$$

where C.C. indicates complex conjugate and φ_0 is a constant. After doing variations with respect to the fields and using partial integrations with surface terms discarded we obtain the dynamic equations:

$$\Box_y\Box_z\,\varphi_2 - 2\varphi_1(\lambda\varphi_2^*\varphi_2 - m^2) = 0 \ \text{Variation WRT}\ \varphi_1 \tag{11.3a}$$

---

[85] Appendix C of Blaha (2016f) presents a first quantized PseudoQuantum theory CQ Mechanics that embodies both classical and quantum theory. **That theory is the non-relativistic quantum mechanics limit of the PseudoFermion theory developed here.** CQ Mechanics has two sets of coordinates that combine to create a generalization of conventional Quantum Mechanics. It has applications in a generalized Feynman path integral formalism, a generalized Schrödinger equation, a generalized Boltzmann equation, the Fokker-Planck equation, a generalized approach to quantum and classical chaos, and to quantum entanglement as well as semi-quantum entanglement. Our "Pseudo" formalisms apply to both Quantum Field Theory and Quantum Mechanics as well as the path integrals, the Fokker-Planck equation and the Boltzmann equation. In these applications there is a clear almost continuous transition from the quantum to the classical sectors.

[86] The choice of lower order derivative terms leads to unsatisfactory results. The fourth order derivatives in eq. 8.3 raise the issue of a linear potential. See S. Blaha, Phys. Rev. **D10**, 4268 (1974).

$$\Box_y\Box_z\,\varphi_1 - 2\varphi_2(\lambda\varphi_1{}^*\varphi_1 - m^2) = 0 \quad \text{Variation WRT } \varphi_2 \quad (11.3b)$$

with $m = \sqrt{\lambda}\varphi_0$.

If we assume the subsidiary condition:

$$\Box_z\,\varphi_2 = m^2\varphi_1 \quad (11.4)$$

we obtain from eq. 11.3a:

$$\Box_y\varphi_1 - 2\varphi_1(\lambda\varphi_2{}^*\varphi_2 - m^2)/m^2 = 0 \quad \text{bradyon equation} \quad (11.5)$$

Similarly the subsidiary condition:

$$\Box_y\,\varphi_2 = -m^2\varphi_1 \quad (11.6)$$

leads (by eq. 11.3a) to

$$\Box_z\varphi_1 + 2\varphi_1(\lambda\varphi_2{}^*\varphi_2 - m^2)/m^2 = 0 \quad \text{tachyon equation} \quad (11.7)$$

Correspondingly eq. 11.2b gives (by $1 \leftrightarrow 2$ symmetry)

$$\Box_z\,\varphi_1 = m^2\varphi_2 \quad (11.8)$$

$$\Box_y\varphi_2 - 2\varphi_2(\lambda\varphi_1{}^*\varphi_1 - m^2)/m^2 = 0 \quad \text{bradyon equation} \quad (11.9)$$

Similarly the subsidiary condition is:

$$\Box_y\,\varphi_1 = -m^2\varphi_2 \quad (11.10)$$

$$\Box_z\varphi_2 + 2\varphi_2(\lambda\varphi_1{}^*\varphi_1 - m^2)/m^2 = 0 \quad \text{tachyon equation} \quad (11.11)$$

Going to the free field case the above equations display the bradyon y parts and tachyon z parts of the Higgsian fields. The process of using these fields to further describe the Higgs Mechanism proceeds in a manner similar to the familiar way.

11.2 Free U(1) Duplex Vector PseudoBoson Formalism

This formalism is for electromagnetic-like vector fields with a gauge transformation. The Bradyon-Tachyon distinction is irrelevant in this case since the fields are massless. The y coordinates are formally bradyonic and the z coordinates are formally tachyonic.

The duplex vector fields are $A_i{}^\mu(y, z)$ for $i = 1, 2$. The y part is bradyonic and the z part is tachyonic. The field strengths are

$$\begin{aligned}
F^{1y}{}_{1\mu\nu\alpha} &= \partial_{z\alpha}(\partial_{y\nu}A_{1\mu} - \partial_{y\mu}A_{1\nu}) \\
F^{2y}{}_{1\mu\nu\alpha} &= \partial_{z\alpha}(\partial_{y\nu}A_{2\mu} - \partial_{y\mu}A_{2\nu})
\end{aligned} \quad (11.12)$$

$$\begin{aligned}
F^{1z}{}_{1\mu\nu\alpha} &= \partial_{y\alpha}(\partial_{z\nu}A_{1\mu} - \partial_{z\mu}A_{1\nu}) \\
F^{2z}{}_{1\mu\nu\alpha} &= \partial_{y\alpha}(\partial_{z\nu}A_{2\mu} - \partial_{z\mu}A_{2\nu})
\end{aligned}$$

due to the need for gauge invariance seen below. The additional derivatives introduced in eq. 11.12 are needed to preserve y – z symmetry in the equal time commutators of eq. 11.20. Thus eq. 11.12, and the complete duplex vector PseudoBoson Lagrangian, have gauge invariance separately in the y and z coordinates:

y coordinates

$$A_{i\mu} \rightarrow A_{i\mu} + \partial_{y\mu}\Lambda(y)$$

z coordinates

$$A_{i\mu} \rightarrow A_{i\mu} + \partial_{z\mu}\Lambda(z)$$

for i = 1,2.

The free duplex vector PseudoBoson Lagrangian with Higgs Mechanism implemented becomes

$$\mathscr{L} = - \tfrac{1}{2}\,(F^{1y}{}_{\mu\nu\alpha}\, F^{2y\,\mu\nu\alpha} - F^{1z}{}_{\mu\nu\alpha}\, F^{2z\,\mu\nu\alpha}\,) - 2e^2\,\varphi_0{}^2\, A_{1\mu}\, A_2{}^{\mu} \qquad (11.13)$$

where e is the coupling constant. After doing variations with respect to the fields and using partial integrations with surface terms discarded we obtain the dynamic equations:

$$\Box_y{}^2\, A_1{}^{\mu} + 2e^2\,\varphi_0{}^2 A_1{}^{\mu} = \Box_z{}^2\, A_1{}^{\mu} - 2e^2\,\varphi_0{}^2 A_1{}^{\mu} = 0 \qquad (11.14)$$

$$\Box_y{}^2\, A_2{}^{\mu} + 2e^2\,\varphi_0{}^2 A_2{}^{\mu} = \Box_z{}^2\, A_2{}^{\mu} - 2e^2\,\varphi_0{}^2 A_2{}^{\mu} = 0 \qquad (11.15)$$

where $\Box_a{}^2 = (\partial_{a\mu}\partial_a{}^{\mu})^2$ for a = y, z in the radiation gauge where $A_1{}^0 = A_2{}^0 = 0$ and

$$\partial_{yi}\, A_2{}^i = \partial_{zi}\, A_2{}^i = 0 \qquad (11.16)$$

where the y and z subscripts indicate derivatives with respect to y and z respectively. Note eqs. 11.14 – 11.15 have bradyon y coordinates and tachyon z coordinates.

The derivation of features parallels that of eqs. 10.17 – 10.23.

11.3 Non-Abelian Duplex Vector PseudoBoson Formalism

Non-Abelian PseudoQuantum gauge field formalisms[87] were developed by this author in the 1970s. These formalisms implemented quark confinement with a linear potential. The non-Abelian vector PseudoBoson theory would also support a quark confinement model with a linear confining potential.

This formalism for the non-Abelian duplex gauge vector fields of a vector gauge boson has the form $A_{i\mu}(y, z)$ for i = 1, 2. The y part of each field will be seen to be bradyonic and the z part of each field will be tachyonic. The field strengths are

$$F^{1y}{}_{1\mu\nu\alpha} = \partial_{z\alpha}\,(\partial_{yv}\, A_{1\mu} - \partial_{y\mu}\, A_{1v} + gA_{1\mu} \times A_{1v}) \qquad (11.17)$$

$$F^{2y}{}_{1\mu\nu\alpha} = \partial_{z\alpha}\,(\partial_{yv}\, A_{2\mu} - \partial_{y\mu}\, A_{2v} + gA_{1\mu} \times A_{2v} - gA_{1v} \times A_{2\mu})$$

$$F^{1z}{}_{1\mu\nu\alpha} = \partial_{y\alpha}\,(\partial_{zv}\, A_{1\mu} - \partial_{z\mu}\, A_{1v} + gA_{1\mu} \times A_{1v})$$

---

[87] S. Blaha, Phys. Rev. **D10**, 4268 (1974); Phys. Rev. **D11**, 2921 (1975); Il Nuovo Cimento **49A**, 35 (1979).

$$F^{2z}{}_{1\mu\nu\alpha} = \partial_{y\alpha}\,(\partial_{zv}\,A_{2\mu} - \partial_{z\mu}\,A_{2v} + gA_{1\mu} \times A_{2v} - gA_{1v} \times A_{2\mu})$$

where $\times$ indicates a commutator. The free duplex vector PseudoBoson Lagrangian[88] with Higgs Mechanism implemented[89] becomes

$$\mathscr{L} = \tfrac{1}{2}\,(F^{1y}{}_{\mu\nu\alpha} \cdot F^{2y\,\mu\nu\alpha} - F^{1z}{}_{\mu\nu\alpha} \cdot F^{2z\,\mu\nu\alpha}) - 2e^2\,\varphi_0{}^2\,A_{1\mu} \cdot A_2{}^\mu \qquad (11.18)$$

where $\cdot$ indicates matrix multiplication.

The free part of $\mathscr{L}$ (truncated to quadratic terms in $A_{1\mu}$ and $A_{2\mu}$) leads to dynamic equations similar to eqs. 11.14 and 11.15 exhibiting the bradyonic and tachyonic parts of the fields.

The derivation of features parallels that of eqs. $10.17 - 10.23$.

---

[88] S. Blaha, Phys. Rev. **D11**, 2921 (1975) develops the formalism for this Lagrangian, which is eq. 17 in that paper, for the case where $z = y$. Eq. 98 of the paper has the quartic form $\Box\nabla^2\,A_1{}^0 = g\lambda^2 J^0$, which indicates a k^{-4} gauge field propagator and thereby a linear confining potential.

[89] In this model calculation we treat all vector bosons as acquiring the same Higgs mass. The generalization to realistic cases is direct.

12. Complex Momentum Conservation

The introduction of tachyons into Quantum Field Theory, and the subsequent definition of interacting combined bradyon-tachyon fields raises the question of momentum conservation. One might have separate momentum conservation for the real and imaginary momenta. Then one would have particle interactions that do not have exchanges of momentum between real and imaginary parts. This approach would be viable. But it would prevent transitions between bradyons and tachyons. It would also prevent faster-than-light starship travel.[90] Momentum transitions would appear to be necessary for speedy starships with tachyonic motion capabilities.[91]

For this reason we now generalize baryon-tachyon dynamics to the case of *Conserved total complex momentum*. Within this framework there is a sector with strict real momentum conservation as we have hitherto seen in Nature. There is also a sector with strict imaginary momentum conservation. Only the total sum of real and imaginary parts is truly conserved.

Then it would be possible to have momentum interchange between real and imaginary momentum. Such an exchange would take place through *complex-valued* forces.[92]

We now will consider particle interactions that might generate an effective complex force. This type of interaction has not been found as yet in Nature. We suggest this type of interaction might exist but that it has not been recognized as such. For example we know that Dark Matter exists due to its gravitational effects. But Dark Matter has not been conclusively found to interact with Normal matter. Similarly,

---

[90] Except possibly for exotic methods such as travel through worm holes. These suggested methods are still in the realm of the imaginable but impossible with current or forseeable technology. Other approaches such as ion drives, using solar energies and so are also beyond current technology and make travel to the stars very lengthy and thus not practically feasible. This author wrote several some years ago making these points.

[91] Books ranging back to:
Blaha (2011b): *All the Universe! Faster Than Light Tachyon Quark Starships & Particle Accelerators with the LHC as a Prototype Starship Drive Scientific Edition.*
Blaha (2011c): *From Asynchronous Logic to The Standard Model to Superflight to the Stars.*
Blaha (2012a): *From Asynchronous Logic to The Standard Model to Superflight to the Stars volume 2: Superluminal CP and CPT, U(4) Complex General Relativity and The Standard Model, Complex Vierbein General Relativity, Kinetic Theory, Thermodynamics.*
And so on.

[92] There are no *known complex* forces in Nature. One might think that resonance decay, which uses Breit-Wigner fits with complex momenta, might be based on a complex interaction force. However the author's gambol models of resonance decay show that Breit-Wigner fits are merely suggestive. Resonance decays have no complex forces at work. See Blaha (2023e). Another proposed case of complexity – non-Hermitean edge effects have complex energies, but not complex forces, since these effects are based on real-valued electromagnetism. See Lei Xiao *et al*, Phys. Rev. Lett., **133**, 070801 (2024).

bradyonic-tachyonic interactions have not been found in Nature. But the possibility of their existence remains.[93]

12.1 Bradyon-Tachyon Wave Functions

This section discusses the interchange of bradyon and tachyon momentum in Quantum Field Theory. Its purpose is to support the quantum equivalent of complex valued forces composed of real and imaginary parts. These forces, on a macroscopic scale, could support starships that achieve complex valued velocities.

As we have shown previously, complex valued velocities enable a starship to rise from subluminal velocity, and bypass the $v = c$ barrier by virtue of its complex velocity. We view the magnitude of velocity as existing in a complex plane with the $v = c$ point as a branch point. Going around $v = c$ a starship, by using a combination of real and imaginary velocity, would enable a ship to reach extremely large real-valued velocities after decelerating the imaginary part of the velocity.

We now turn to quantum theory models where the tachyon parts of a set of interacting particles exchanges momentum with the bradyon parts of the set of interacting particles. The total tachyon plus bradyon momentum is conserved while the sum of the tachyon momenta is not conserved, and the sum of the bradyon momenta are also not conserved. This possibility has not yet been observed in Nature.[94]

12.1.1 Tachyon-Bradyon Momentum Interchange

We can define a simple Lagrangian interaction that leads to the interchange of the bradyon and tachyon parts of a fermion

$$m\bar{\psi}_{2BT\beta\alpha}(z, y)\psi_{2TB}{}^{\alpha\beta}(y, z) \tag{12.1}$$

using eq. 7.1

$$\psi_{iTB\alpha\beta}(y, z) = \psi_{iBT\beta\alpha}(z, y) \tag{7.1}$$

for a fermion with bradyon and tachyon mass M. Eq. 12.1 may be viewed as a mass-like term that implements transitions such as those visualized in Fig. 12.3.

This interaction interchanges the y and z coordinates parts of a fermion. For example the fermion with a bradyon LAB part and a tachyon interior part has its parts interchanged as shown in the following wave functions:

$$\psi_{iBT\alpha\beta}(y, z) = \underset{s_1\ s_2}{\Sigma\Sigma} \int dp^{r-1}\int dq^{r-1}\ N(p, q)\ [b_{iBT}(p, q, s_1, s_2)u_\alpha(p, s_1)u_{\beta BT}(q, s_2)\exp(-ip\cdot y - iq\cdot z) +$$

$$+ \, d_{iBT}^{\dagger}(p, \, q, s1, s2) v_{\alpha}(p, s1) v_{\beta BT}(q, \, s2) \exp(ip \cdot y + iq \cdot z)] \qquad (5.14)$$

becomes

$$\psi_{iTB\beta\alpha}(z, \, y) = \Sigma\Sigma \int dp^{r-1} \int dq^{r-1} \, N(p, \, q) \, [b_{iBT}(p, \, q, s_1, s_2) u_{\beta}(p, s_1) u_{\alpha BT}(q, \, s_2) \, \exp(-ip \cdot z - iq \cdot y) +$$
$$_{s_1 \; s_2}$$
$$+ \, d_{iBT}^{\dagger}(p, \, q, s1, s2) v_{\beta}(p, s1) v_{\alpha BT}(q, \, s2) \exp(ip \cdot z + iq \cdot y)] \qquad (12.2)$$

Figure 12.1. Feynman-like diagram for fermion transitions between a bradyon part with momentum p and a tachyon part of momentum q to a bradyon part with momentum q and a tachyon part of momentum p and vice versa.

The interchange does not conserve the momentum of each part separately. The LAB momentum goes from p to q in the wave function. The interior part momentum goes from q to p in the wave function. The total momentum of the Fourier components of the parts p + q is conserved. The exponential term $ip \cdot y + iq \cdot z$ may be expressed as

$$ip \cdot y + iq \cdot z = i(p + q) \cdot y_1 + i(p - q) \cdot y_2 \qquad (12.3)$$

using $y = y_1 + y_2$ and $z = y_1 - y_2$. Thus p + q may be viewed as the total momentum.[95] The p − q part is an interior relative momentum.

12.2 Complex Momentum Conservation in Duplex Particle Interactions

One can consider more realistic fermion interactions in which only total complex momentum is conserved. Consider an interaction term such as

$$g \, [\overline{\psi}(y, \, z)\psi(y, \, z)]^2 \qquad (12.4)$$

It is a duplex Quantum Field Theory interaction that enables momentum transfers between real and imaginary momentum. See Figs. 12.2 and 12.3 for examples.

They consider cases of a composite bradyon-tachyon interacting with a composite tachyon-bradyon to create:

1) Another pair of composite bradyon-tachyons where the LAB momentum is bradyonic (Fig. 12.2); and

2) Another pair of particles, one of which is pure bradyon, and the other is pure tachyon. (Fig. 12.3) Tachyon particle creation!

---

[95] The conservation of total *complex* momentum and angular momentum may be derived in Quantum Field Theory in a manner similar to the conserved real-valued quantities.

12.3 QuadPlex Case

Quadplex fermions have similar, but more complicated, sets of interactions. They also have total momentum conservation for the sum of complex momenta parts but not for the real and imaginary parts separately.

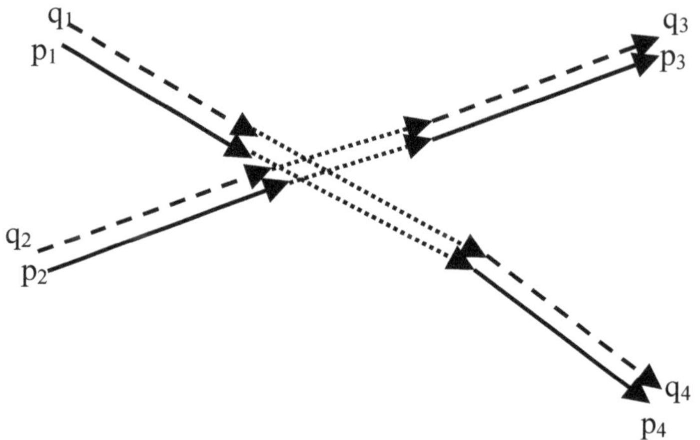

Figure 12.2. Interaction of a composite bradyon-tachyon with a composite tachyon-bradyon. Bradyon momenta designated with p and tachyon momenta designated with q. The p_i momentum in a pair is the LAB momentum. The q_i momentum in a pair is the internal momentum. Lines that cross in the interaction region are not interacting. The incoming particles exchange parts to form outgoing particles. This is but one of many possible configurations for the interactions of the type of eq. 12.4.

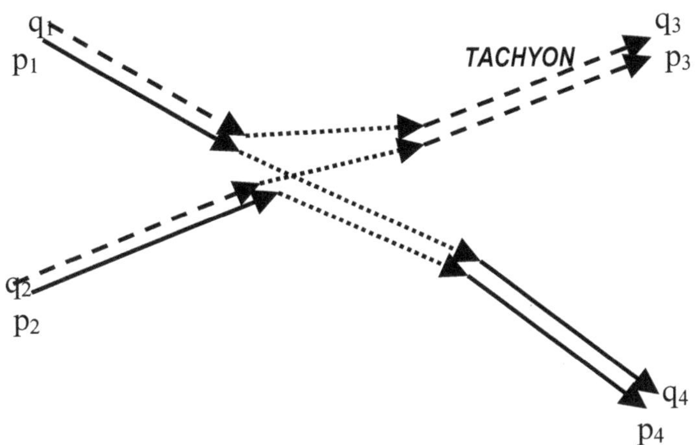

Figure 12.3. Interaction of a composite bradyon-tachyon with a composite tachyon-bradyon. Lines that cross in the interaction region are not interacting. This diagram shows the creation of a pure tachyon. LAB plus LAB bradyon fermions → LAB bradyon fermion plus LAB tachyon fermion.

12.3 Momentum Conservation

The sum of real and imaginary duplex momentum parts in an interaction is conserved. Thus for Figs. 12.2 and 12.3 we find

$$q_1 + q_2 + p_1 + p_2 = q_3 + q_4 + p_3 + p_4 \tag{12.5}$$

However the separate sums of the real and imaginary parts are not conserved in Fig. 12.3:

$$q_1 + q_2 \neq q_3 + q_4 \tag{12.6}$$
$$p_1 + p_2 \neq p_3 + p_4 \tag{12.7}$$

12.3.1 Example of Separate Momentum Conservation

Interactions with separate momentum conservation for the LAB momentum and the non-LAB momentum appear in Fig. 12.4. The conserved momenta are

$$q_1 + q_2 = q_3 + q_4 \tag{12.8}$$
$$p_1 + p_2 = p_3 + p_4 \tag{12.9}$$

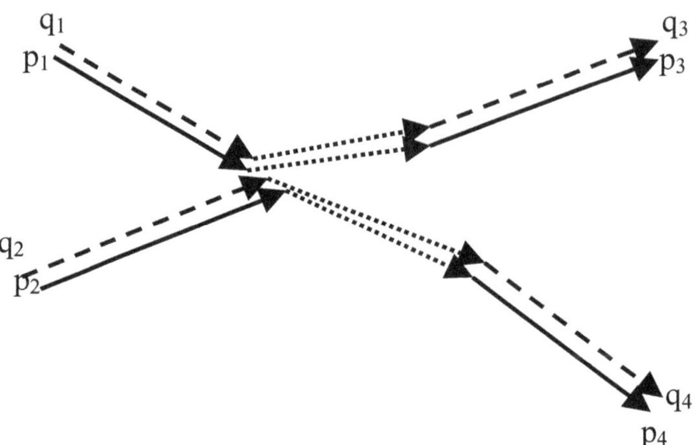

Figure 12.4. Interaction of a composite bradyon-tachyon with a composite tachyon-bradyon. Bradyon momenta designated with p and tachyon momenta designated with q. The p_i momentum in a pair is the LAB momentum. The q_i momentum in a pair is the internal momentum. Lines that cross in the interaction region are not interacting. This diagram shows separate conservation of the LAB momentum and separate conservation of the imaginary momentum.

Interactions with separate momentum conservation for each of the four parts of quadplex fermions may be defined. *Thus we have momentum conservation possible for the LAB parts for both duplex and quadplex fermions as seen in conventional kinematics. The LAB part and the other parts support total momentum conservation in all cases as seen above.*

10.4 Complex Force Law

The preceding sections consider cases of individual particle interactions. They generalize to the case of macroscopic masses of particles. They open the door to the possibility of superluminal motion for starships.

The essential role in superluminal macroscopic motion is that of tachyon propulsion through the generation of tachyons in a manner similar to Fig. 12.3 in macroscopic quantities.

A tachyon engine remains to be developed. First we need duplex or quadplex fermions that can undergo reactions similar to Fig. 12.3 on a macroscopic scale. Secondly we need a concentrated[96] source of energy that would power starships to superluminal complex velocities. Atomic energy and/or fusion might not be enough. A

---

[96] The thrust energy per unit mass must be sufficiently large taking account of both the fuel mass and the mass of the container/engine.

type of vacuum energy that we have considered in our recent books might suffice. We will consider these topics in future studies.

REFERENCES

Akhiezer, N. I., Frink, A. H. (tr), 1962, *The Calculus of Variations* (Blaisdell Publishing, New York, 1962).

Bjorken, J. D., Drell, S. D., 1964, *Relativistic Quantum Mechanics* (McGraw-Hill, New York, 1965).

Bjorken, J. D., Drell, S. D., 1965, *Relativistic Quantum Fields* (McGraw-Hill, New York, 1965).

Blaha, S., 1995, *C++ for Professional Programming* (International Thomson Publishing, Boston, 1995).

________, 1998, *Cosmos and Consciousness* (Pingree-Hill Publishing, Auburn, NH, 1998 and 2002).

________, 2002, *A Finite Unified Quantum Field Theory of the Elementary Particle Standard Model and Quantum Gravity Based on New Quantum Dimensions™ & a New Paradigm in the Calculus of Variations* (Pingree-Hill Publishing, Auburn, NH, 2002).

________, 2004, *Quantum Big Bang Cosmology: Complex Space-time General Relativity, Quantum Coordinates™ Dodecahedral Universe, Inflation, and New Spin 0, ½, 1 & 2 Tachyons & Imagyons* (Pingree-Hill Publishing, Auburn, NH, 2004).

________, 2005a, *Quantum Theory of the Third Kind: A New Type of Divergence-free Quantum Field Theory Supporting a Unified Standard Model of Elementary Particles and Quantum Gravity based on a New Method in the Calculus of Variations* (Pingree-Hill Publishing, Auburn, NH, 2005).

________, 2005b, *The Metatheory of Physics Theories, and the Theory of Everything as a Quantum Computer Language* (Pingree-Hill Publishing, Auburn, NH, 2005).

________, 2005c, *The Equivalence of Elementary Particle Theories and Computer Languages: Quantum Computers, Turing Machines, Standard Model, Superstring Theory, and a Proof that Gödel's Theorem Implies Nature Must Be Quantum* (Pingree-Hill Publishing, Auburn, NH, 2005).

________, 2006a, *The Foundation of the Forces of Nature* (Pingree-Hill Publishing, Auburn, NH, 2006).

________, 2006b, *A Derivation of ElectroWeak Theory based on an Extension of Special Relativity; Black Hole Tachyons; & Tachyons of Any Spin.* (Pingree-Hill Publishing, Auburn, NH, 2006).

________, 2007a, *Physics Beyond the Light Barrier: The Source of Parity Violation, Tachyons, and A Derivation of Standard Model Features* (Pingree-Hill Publishing, Auburn, NH, 2007).

________, 2007b, *The Origin of the Standard Model: The Genesis of Four Quark and Lepton Species, Parity Violation, the ElectroWeak Sector, Color SU(3), Three Visible Generations of Fermions, and One Generation of Dark Matter with Dark Energy* (Pingree-Hill Publishing, Auburn, NH, 2007).

________, 2008a, *A Direct Derivation of the Form of the Standard Model From GL(16)* (Pingree-Hill Publishing, Auburn, NH, 2008).

________, 2008b, *A Complete Derivation of the Form of the Standard Model With a New Method to Generate Particle Masses Second Edition* (Pingree-Hill Publishing, Auburn, NH, 2008)

________, 2009, *The Algebra of Thought & Reality: The Mathematical Basis for Plato's Theory of Ideas, and Reality Extended to Include A Priori Observers and Space-Time Second Edition* (Pingree-Hill Publishing, Auburn, NH, 2009).

______, 2010a, *Operator Metaphysics: A New Metaphysics Based on a New Operator Logic and a New Quantum Operator Logic that Lead to a Mathematical Basis for Plato's Theory of Ideas and Reality* (Pingree-Hill Publishing, Auburn, NH, 2010).

______, 2010b, *The Standard Model's Form Derived from Operator Logic, Superluminal Transformations and GL(16)* (Pingree-Hill Publishing, Auburn, NH, 2010).

______, 2010c, *SuperCivilizations: Civilizations as Superorganisms* (McMann-Fisher Publishing, Auburn, NH, 2010).

______, 2011a, *21*^st *Century Natural Philosophy Of Ultimate Physical Reality* (McMann-Fisher Publishing, Auburn, NH, 2011).

______, 2011b, *All the Universe! Faster Than Light Tachyon Quark Starships & Particle Accelerators with the LHC as a Prototype Starship Drive Scientific Edition* (Pingree-Hill Publishing, Auburn, NH, 2011).

______, 2011c, *From Asynchronous Logic to The Standard Model to Superflight to the Stars* (Blaha Research, Auburn, NH, 2011).

______, 2012a, *From Asynchronous Logic to The Standard Model to Superflight to the Stars volume 2: Superluminal CP and CPT, U(4) Complex General Relativity and The Standard Model, Complex Vierbein General Relativity, Kinetic Theory, Thermodynamics* (Blaha Research, Auburn, NH, 2012).

______, 2012b, *Standard Model Symmetries, And Four And Sixteen Dimension Complex Relativity; The Origin Of Higgs Mass Terms* (Blaha Reasearch, Auburn, NH, 2012).

______, 2013a, *Multi-Stage Space Guns, Micro-Pulse Nuclear Rockets, and Faster-Than-Light Quark-Gluon Ion Drive Starships* (Blaha Research, Auburn, NH, 2013).

______, 2013b, *The Bridge to Dark Matter; A New Sibling Universe; Dark Energy; Inflatons; Quantum Big Bang; Superluminal Physics; An Extended Standard Model Based on Geometry* (Blaha Reasearch, Auburn, NH, 2013).

______, 2014a, *Universes and Megaverses: From a New Standard Model to a Physical Megaverse; The Big Bang; Our Sibling Universe's Wormhole; Origin of the Cosmological Constant, Spatial Asymmetry of the Universe, and its Web of Galaxies; A Baryonic Field between Universes and Particles; Megaverse Extended Wheeler-DeWitt Equation* (Blaha Reasearch, Auburn, NH, 2014).

______, 2014b, *All the Megaverse! Starships Exploring the Endless Universes of the Cosmos Using the Baryonic Force* (Blaha Research, Auburn, NH, 2014).

______, 2014c, *All the Megaverse! II Between Megaverse Universes: Quantum Entanglement Explained by the Megaverse Coherent Baryonic Radiation Devices – PHASERs Neutron Star Megaverse Slingshot Dynamics Spiritual and UFO Events, and the Megaverse Microscopic Entry into the Megaverse* (Blaha Research, Auburn, NH, 2014).

______, 2015a, *PHYSICS IS LOGIC PAINTED ON THE VOID: Origin of Bare Masses and The Standard Model in Logic, U(4) Origin of the Generations, Normal and Dark Baryonic Forces, Dark Matter, Dark Energy, The Big Bang, Complex General Relativity, A Megaverse of Universe Particles* (Blaha Research, Auburn, NH, 2015).

______, 2015b, *PHYSICS IS LOGIC Part II: The Theory of Everything, The Megaverse Theory of Everything, U(4)⊗U(4) Grand Unified Theory (GUT), Inertial Mass = Gravitational Mass, Unified Extended Standard Model and a New Complex General Relativity with Higgs Particles, Generation Group Higgs Particles* (Blaha Research, Auburn, NH, 2015).

______, 2015c, *The Origin of Higgs ("God") Particles and the Higgs Mechanism: Physics is Logic III, Beyond Higgs – A Revamped Theory With a Local Arrow of Time, The Theory of Everything Enhanced, Why Inertial Frames are Special, Universes of the Mind* (Blaha Research, Auburn, NH, 2015).

______, 2015d, *The Origin of the Eight Coupling Constants of The Theory of Everything: U(8) Grand Unified Theory of Everything (GUTE), S^8 Coupling Constant Symmetry, Space-Time Dependent Coupling Constants, Big Bang Vacuum Coupling Constants, Physics is Logic IV* (Blaha Research, Auburn, NH, 2015).

______, 2016a, *New Types of Dark Matter, Big Bang Equipartition, and A New U(4) Symmetry in the Theory of Everything: Equipartition Principle for Fermions, Matter is 83.33% Dark, Penetrating the Veil of the Big Bang, Explicit QFT Quark Confinement and Charmonium, Physics is Logic V* (Blaha Research, Auburn, NH, 2016).

______, 2016b, *The Periodic Table of the 192 Quarks and Leptons in The Theory of Everything: The U(4) Layer Group, Physics is Logic VI* (Blaha Research, Auburn, NH, 2016).

______, 2016c, *New Boson Quantum Field Theory, Dark Matter Dynamics, Dark Matter Fermion Layer Mixing, Genesis of Higgs Particles, New Layer Higgs Masses, Higgs Coupling Constants, Non-Abelian Higgs Gauge Fields, Physics is Logic VII* (Blaha Research, Auburn, NH, 2016).

______, 2016d, *Unification of the Strong Interactions and Gravitation: Quark Confinement Linked to Modified Short-Distance Gravity; Physics is Logic VIII* (Blaha Research, Auburn, NH, 2016).

______, 2016e, *MoND: Unification of the Strong Interactions and Gravitation II, Quark Confinement Linked to Large-Scale Gravity, Physics is Logic IX* (Blaha Research, Auburn, NH, 2016).

______, 2016f, *CQ Mechanics: A Unification of Quantum & Classical Mechanics, Quantum/Semi-Classical Entanglement, Quantum/Classical Path Integrals, Quantum/Classical Chaos* (Blaha Research, Auburn, NH, 2016).

______, 2016g, *GEMS Unified Gravity, ElectroMagnetic and Strong Interactions: Manifest Quark Confinement, A Solution for the Proton Spin Puzzle, Modified Gravity on the Galactic Scale* (Pingree Hill Publishing, Auburn, NH, 2016).

______, 2016h, *Unification of the Seven Boson Interactions based on the Riemann-Christoffel Curvature Tensor* (Pingree Hill Publishing, Auburn, NH, 2016).

______, 2017a, *Unification of the Eleven Boson Interactions based on 'Rotations of Interactions'* (Pingree Hill Publishing, Auburn, NH, 2017).

______, 2017b, *The Origin of Fermions and Bosons, and Their Unification* (Pingree Hill Publishing, Auburn, NH, 2017).

______, 2017c, *Megaverse: The Universe of Universes* (Pingree Hill Publishing, Auburn, NH, 2017).

______, 2017d, *SuperSymmetry and the Unified SuperStandard Model* (Pingree Hill Publishing, Auburn, NH, 2017).

______, 2017e, *From Qubits to the Unified SuperStandard Model with Embedded SuperStrings: A Derivation* (Pingree Hill Publishing, Auburn, NH, 2017).

______, 2017f, *The Unified SuperStandard Model in Our Universe and the Megaverse: Quarks, … ,* (Pingree Hill Publishing, Auburn, NH, 2017).

______, 2018a, *The Unified SuperStandard Model and the Megaverse SECOND EDITION A Deeper Theory based on a New Particle Functional Space that Explicates Quantum Entanglement Spookiness (Volume 1)* (Pingree Hill Publishing, Auburn, NH, 2018).

______, 2018b, *Cosmos Creation: The Unified SuperStandard Model, Volume 2, SECOND EDITION* (Pingree Hill Publishing, Auburn, NH, 2018).

______, 2018c, *God Theory* (Pingree Hill Publishing, Auburn, NH, 2018).

______, 2018d, *Immortal Eye: God Theory: Second Edition* (Pingree Hill Publishing, Auburn, NH, 2018).

______, 2018e, *Unification of God Theory and Unified SuperStandard Model THIRD EDITION* (Pingree Hill Publishing, Auburn, NH, 2018).

______, 2019a, *Calculation of: QED α = 1/137, and Other Coupling Constants of the Unified SuperStandard Theory* (Pingree Hill Publishing, Auburn, NH, 2019).

______, 2019b, *Coupling Constants of the Unified SuperStandard Theory SECOND EDITION* (Pingree Hill Publishing, Auburn, NH, 2019).

______, 2019c, *New Hybrid Quantum Big_Bang–Megaverse_Driven Universe with a Finite Big Bang and an Increasing Hubble Constant* (Pingree Hill Publishing, Auburn, NH, 2019).

______, 2019d, *The Universe, The Electron and The Vacuum* (Pingree Hill Publishing, Auburn, NH, 2019).

______, 2019e, *Quantum Big Bang – Quantum Vacuum Universes (Particles)* (Pingree Hill Publishing, Auburn, NH, 2019).

______, 2019f, *The Exact QED Calculation of the Fine Structure Constant Implies ALL 4D Universes have the Same Physics/Life Prospects* (Pingree Hill Publishing, Auburn, NH, 2019).

______, 2019g, *Unified SuperStandard Theory and the SuperUniverse Model: The Foundation of Science* (Pingree Hill Publishing, Auburn, NH, 2019).

______, 2020a, *Quaternion Unified SuperStandard Theory (The QUeST) and Megaverse Octonion SuperStandard Theory (MOST)* (Pingree Hill Publishing, Auburn, NH, 2020).

______, 2020b, *United Universes Quaternion Universe - Octonion Megaverse* (Pingree Hill Publishing, Auburn, NH, 2020).

______, 2020c, *Unified SuperStandard Theories for Quaternion Universes & The Octonion Megaverse* (Pingree Hill Publishing, Auburn, NH, 2020).

______, 2020d, *The Essence of Eternity: Quaternion & Octonion SuperStandard Theories* (Pingree Hill Publishing, Auburn, NH, 2020).

______, 2020e, *The Essence of Eternity II* (Pingree Hill Publishing, Auburn, NH, 2020).

______, 2020f, *A Very Conscious Universe* (Pingree Hill Publishing, Auburn, NH, 2020).

______, 2020g, *Hypercomplex Universe* (Pingree Hill Publishing, Auburn, NH, 2020).

______, 2020h, *Beneath the Quaternion Universe* (Pingree Hill Publishing, Auburn, NH, 2020).

______, 2020i, *Why is the Universe Real? From Quaternion & Octonion to Real Coordinates* (Pingree Hill Publishing, Auburn, NH, 2020).

______, 2020j, *The Origin of Universes: of Quaternion Unified SuperStandard Theory (QUeST); and of the Octonion Megaverse (UTMOST)* (Pingree Hill Publishing, Auburn, NH, 2020).

______, 2020k, *The Seven Spaces of Creation: Octonion Cosmology* (Pingree Hill Publishing, Auburn, NH, 2020).

________, 2020l, *From Octonion Cosmology to the Unified SuperStandard Theory of Particles* (Pingree Hill Publishing, Auburn, NH, 2020).

________, 2021a, *Pioneering the Cosmos* (Pingree Hill Publishing, Auburn, NH, 2021).

________, 2021b, *Pioneering the Cosmos II* (Pingree Hill Publishing, Auburn, NH, 2021).

________, 2021c, *Beyond Octonion Cosmology* (Pingree Hill Publishing, Auburn, NH, 2021).

________, 2021d, *Universes are Particles* (Pingree Hill Publishing, Auburn, NH, 2021).

________, 2021e, *Octonion-like dna-based life, Universe expansion is decay, Emerging New Physics* (Pingree Hill Publishing, Auburn, NH, 2021).

________, 2021f, *The Science of Creation New Quantum Field Theory of Spaces* (Pingree Hill Publishing, Auburn, NH, 2021).

________, 2021g, *Quantum Space Theory With Application to Octonion Cosmology & Possibly To Fermionic Condensed Matter* (Pingree Hill Publishing, Auburn, NH, 2021).

________, 2021h, *21st Century Natural Philosophy of Octonion Cosmology , and Predestination, Fate, and Free Will* (Pingree Hill Publishing, Auburn, NH, 2021).

________, 2021i, *Beyond Octonion Cosmology II : Origin of the Quantum; A New Generalized Field Theory (GiFT); A Proof of the Spectrum of Universes; Atoms in Higher Universes* (Pingree Hill Publishing, Auburn, NH, 2021).

________, 2021j, *Integration of General Relativity and Quantum Theory: Octonion Cosmology, GiFT, Creation/Annihilation Spaces CASe, Reduction of Spaces to a Few Fermions and Symmetries in Fundamental Frames* (Pingree Hill Publishing, Auburn, NH, 2021).

________, 2022a, *New View of Octonion Cosmology Based on the Unification of General Relativity and Quantum Theory* (Pingree Hill Publishing, Auburn, NH, 2022).

________, 2022b, *The Dust Beneath Hypercomplex Cosmology* (Pingree Hill Publishing, Auburn, NH, 2022).

________, 2022c, *Passing Through Nature to Eternity: ProtoCosmos, HyperCosmos, Unified SuperStandard Theory* (Pingree Hill Publishing, Auburn, NH, 2022).

________, 2022d, *HyperCosmos Fractionation and Fundamental Reference Frame Based Unification: Particle Inner Space Basis of Parton and Dual Resonance Models* (Pingree Hill Publishing, Auburn, NH, 2022).

________, 2022e, *A New UniDimension ProtoCosmos and SuperString F-Theory Relation to the HyperCosmos* (Pingree Hill Publishing, Auburn, NH, 2022).

________, 2022f, *The Cosmic Panorama: ProtoCosmos, HyperCosmos,Unified SuperStandard Theory (UST) Derivation* (Pingree Hill Publishing, Auburn, NH, 2022).

________, 2022g, *Ultimate Origin: ProtoCosmos and HyperCosmos* (Pingree Hill Publishing, Auburn, NH, 2022).

________, 2023a, *UltraUnification and the Generation of the Cosmos* (Pingree Hill Publishing, Auburn, NH, 2023).

________, 2023b, *God and and Cosmos Theory* (Pingree Hill Publishing, Auburn, NH, 2023).

________, 2023c, *A New Completely Geometric SU(8) Cosmos Theory; New PseudoFermion Fields; Fibonacci-like Dimension Arrays; Ramsey Number Approximation* (Pingree Hill Publishing, Auburn, NH, 2023).

________, 2023d, *Newton's Apple is Now the Fermion* (Pingree Hill Publishing, Auburn, NH, 2023).

______, 2023e, *Cosmos Theory: The Sub-Particle Gambol Model* (Pingree Hill Publishing, Auburn, NH, 2023).

______, 2024a, *Cosmos-Universe-Particle-Gambol Theory* (Pingree Hill Publishing, Auburn, NH, 2024).

______, 2024b, *Fractal Cosmos Theory* (Pingree Hill Publishing, Auburn, NH, 2024).

______, 2024c, *Fractal Cosmic Curve: Tensor-Based CosmosTheory* (Pingree Hill Publishing, Auburn, NH, 2024).

______, 2024d, *The Eternal Form of Cosmos Theory* (Pingree Hill Publishing, Auburn, NH, 2024).

______, 2024e, *The Eternal Form of Cosmos Theory Third Edition* (Pingree Hill Publishing, Auburn, NH, 2024).

______, 2024f, *Fundamental Constants of Cosmos Theory and The Standard Model* (Pingree Hill Publishing, Auburn, NH, 2024).

______, 2024g, *Quark, Lepton, W and Z Masses of Cosmos Theory and The Standard Model* (Pingree Hill Publishing, Auburn, NH, 2024).

______, 2024h, *Geometric Cosmos Geometric Universe* (Pingree Hill Publishing, Auburn, NH, 2024).

______, 2024i, *Particles and Universes of Cosmos Theory* (Pingree Hill Publishing, Auburn, NH, 2024).

Eddington, A. S., 1952, *The Mathematical Theory of Relativity* (Cambridge University Press, Cambridge, U.K., 1952).

Fant, Karl M., 2005, *Logically Determined Design: Clockless System Design With NULL Convention Logic* (John Wiley and Sons, Hoboken, NJ, 2005).

Feinberg, G. and Shapiro, R., 1980, *Life Beyond Earth: The Intelligent Earthlings Guide to Life in the Universe* (William Morrow and Company, New York, 1980).

Gelfand, I. M., Fomin, S. V., Silverman, R. A. (tr), 2000, *Calculus of Variations* (Dover Publications, Mineola, NY, 2000).

Giaquinta, M., Modica, G., Souchek, J., 1998, *Cartesian Coordinates in the Calculus of Variations* Volumes I and II (Springer-Verlag, New York, 1998).

Giaquinta, M., Hildebrandt, S., 1996, *Calculus of Variations* Volumes I and II (Springer-Verlag, New York, 1996).

Gradshteyn, I. S. and Ryzhik, I. M., 1965, *Table of Integrals, Series, and Products* (Academic Press, New York, 1965).

Heitler, W., 1954, *The Quantum Theory of Radiation* (Claendon Press, Oxford, UK, 1954).

Huang, Kerson, 1992, *Quarks, Leptons & Gauge Fields 2nd Edition* (World Scientific Publishing Company, Singapore, 1992).

Jost, J., Li-Jost, X., 1998, *Calculus of Variations* (Cambridge University Press, New York, 1998).

Kaku, Michio, 1993, *Quantum Field Theory*, (Oxford University Press, New York, 1993).

Kirk, G. S. and Raven, J. E., 1962, *The Presocratic Philosophers* (Cambridge University Press, New York, 1962).

Landau, L. D. and Lifshitz, E. M., 1987, *Fluid Mechanics 2nd Edition*, (Pergamon Press, Elmsford, NY, 1987).

Rescher, N., 1967, *The Philosophy of Leibniz* (Prentice-Hall, Englewood Cliffs, NJ, 1967).

Riesz, Frigyes and Sz.-Nagy, Béla, 1990, *Functional Analysis* (Dover Publications, New York, 1990).

Sakurai, J. J., 1964, *Invariance Principles and Elementary Particles* (Princeton University Press, Princeton, NJ, 1964).

Weinberg, S., 1972, *Gravitation and Cosmology* (John Wiley and Sons, New York, 1972).

Weinberg, S., 1995, *The Quantum Theory of Fields Volume I* (Cambridge University Press, New York, 1995).

INDEX

Stephen Blaha is a well-known Physicist and Man of Letters with interests in Science, Society and civilization, the Arts, and Technology. He had an Alfred P. Sloan Foundation scholarship in college. He received his Ph.D. in Physics from Rockefeller University. He has served on the faculties of several major universities. He was also a Member of the Technical Staff at Bell Laboratories, a manager at the Boston Globe Newspaper, a Director at Wang Laboratories, and President of Blaha Software Inc. and of Janus Associates Inc. (NH).

Among other achievements he was a co-discoverer of the "r potential" for heavy quark binding developing the first (and still the only demonstrable) non-Aeolian gauge theory with an "r" potential; first suggested the existence of topological structures in superfluid He-3; first proposed Yang-Mills theories would appear in condensed matter phenomena with non-scalar order parameters; first developed a grammar-based formalism for quantum computers and applied it to elementary particle theories; first developed a new form of quantum field theory without divergences (thus solving a major 60 year old problem that enabled a unified theory of the Standard Model and Quantum Gravity without divergences to be developed); first developed a formulation of complex General Relativity based on analytic continuation from real space-time; first developed a generalized non-homogeneous Robertson-Walker metric that enabled a quantum theory of the Big Bang to be developed without singularities at t = 0; first generalized Cauchy's theorem and Gauss' theorem to complex, curved multi-dimensional spaces; received Honorable Mention in the Gravity Research Foundation Essay Competition in 1978; first developed a physically acceptable theory of faster-than-light particles; first derived a composition of extremums method in the Calculus of Variations; first quantitatively suggested that inflationary periods in the history of the universe were not needed; first proved Gödel's Theorem implies Nature must be quantum; provided a new alternative to the Higgs Mechanism, and Higgs particles, to generate masses; first showed how to resolve logical paradoxes including Gödel's Undecidability Theorem by developing Operator Logic and Quantum Operator Logic; first developed a quantitative harmonic oscillator-like model of the life cycle, and interactions, of civilizations; first showed how equations describing superorganisms also apply to civilizations. A recent book shows his theory applies successfully to the past 14 years of history and to *new* archaeological data on Andean and Mayan civilizations as well as Early Anatolian and Egyptian civilizations.

He first developed an axiomatic derivation of the form of The Standard Model from geometry – space-time properties – The Unified SuperStandard Model. It unifies all the known forces of Nature. It also has a Dark Matter sector that includes a Dark ElectroWeak sector with Dark doublets and Dark gauge interactions. It uses quantum coordinates to remove infinities that crop up in most interacting quantum field theories and additionally to remove the infinities that appear in the Big Bang and generate inflationary growth of the universe. It shows gravity has a MOND-like form without sacrificing Newton's Laws. It relates the interactions of the MOND-like sector of gravity with the r-potential of Quark Confinement. The axioms of the theory lead to the question of their origin. We suggest in the preceding edition of this book it can be attributed to an entity with God-like properties. We explore these properties in "God Theory" and show they predict that the Cosmos exists forever although individual universes (or incarnations of our universe) "come and go." Several other important results emerge from God Theory such a functionally triune God. The Unified SuperStandard Theory has many other important parts described in the Current Edition of *The Unified SuperStandard Theory* and expanded in subsequent volumes.

Blaha has had a major impact on a succession of elementary particle theories: his Ph.D. thesis (1970), and papers, showed that quantum field theory calculations to all orders in ladder approximations could not give scaling deep inelastic electron-nucleon scattering. He later showed the eigenvalue equation for the fine structure constant α in Johnson-Baker-Willey QED had a zero at $\alpha = 1$ not 1/137 by solving the Schwinger-Dyson equations to all orders in an approximation that agreed with exact results to 4th order in α thus ending interest in this theory. In 1979 at Prof. Ken Johnson's (MIT) suggestion he calculated the proton-neutron mass difference in the MIT bag model and found the result had the wrong sign reducing interest in the bag model. These results all appear in Physical Review papers. In the 2000's he repeatedly pointed out the shortcomings of SuperString theory and showed that The Standard Model's form could be derived from space-time geometry by an extension of Lorentz transformations to faster than light transformations. This deeper space-time basis greatly increases the possibility that it is part of THE fundamental theory. Recently, Blaha showed that the Weak interactions differed significantly from the Strong, electromagnetic and gravitation interactions in important respects while these interactions had similar features, and suggested that ElectroWeak theory, which is essentially a glued union of the Weak interactions and Electromagnetism, possibly modulo unknown Higgs particle features, be replaced by a unified theory of the other interactions combined with a stand-alone Weak interaction theory. Blaha also showed that, if Charmonium calculations are taken seriously, the Strong interaction coupling constant is only a factor of five larger than the electromagnetic coupling constant, and thus Strong interaction perturbation theory would make sense and yield physically meaningful results.

In graduate school (1965-71) he wrote substantial papers in elementary particles and group theory: The Inelastic E- P Structure Functions in a Gluon Model. Phys. Lett. B40:501-502,1972; Deep-Inelastic E-P Structure Functions In A Ladder Model With Spin 1/2 Nucleons, Phys.Rev. D3:510-523,1971; Continuum Contributions To The Pion Radius, Phys. Rev. 178:2167-2169,1969; Character Analysis of U(N) and SU(N), J. Math. Phys. <u>10</u>, 2156 (1969); and The Calculation of the Irreducible Characters of the Symmetric Group in Terms of the Compound Characters, (Published as Blaha's Lemma in D. E. Knuth's book: *The Art of Computer Programming Vols. 1 – 4*).

In the early 1980's Blaha was also a pioneer in the development of UNIX for financial, scientific and Internet applications: benchmarked UNIX versions showing that block size was critical for UNIX performance, developing financial modeling software, starting database benchmarking comparison studies, developing Internet-like UNIX networking (1982) and developing a hybrid shell programming technique (1982) that was a precursor to the PERL programming language. He was also the manager of the AT&T ten-year future products development database. His work helped lead to commercial UNIX on computers such as Sun Micros, IBM AIX minis, and Apple computers.

In the 1980's he pioneered the development of PC Desktop Publishing on laser printers and was nominated for three "Awards for Technical Excellence" in 1987 by PC Magazine for PC software products that he designed and developed.

Recently he has developed a theory of Megaverses – actual universes of which our universe is one – with quantum particle-like properties based on the Wheeler-DeWitt equation of Quantum Gravity. He has developed a theory of a baryonic force, which had been conjectured many years ago, and estimated the strength of the force based on discrepancies in measurements of the gravitational constant G. This force, operative in D-dimensional space, can be used to escape from our universe in "uniships" which are the equivalent of the faster-than-light starships proposed in the author's earlier books. Thus travel to other universes, as well as to other stars is possible.

Blaha also considered the complexified Wheeler-DeWitt equation and showed that its limitation to real-valued coordinates and metrics generated a Cosmological Constant in the Einstein equations.

The author has also recently written a series of books on the serious problems of the United States and their solution as well as a book on the decline of Mankind that will follow from current social and genetic trends in Mankind.

In the past twenty years Dr. Blaha has written over 80 books on a wide range of topics. Some recent major works are: *From Asynchronous Logic to The Standard Model to Superflight to the Stars, All the Universe!, SuperCivilizations: Civilizations as Superorganisms, America's Future: an Islamic Surge, ISIS, al Qaeda, World Epidemics, Ukraine, Russia-China Pact, US Leadership Crisis, The Rises and Falls of Man – Destiny – 3000 AD: New Support for a Superorganism MACRO-THEORY of CIVILIZATIONS From CURRENT WORLD TRENDS and NEW Peruvian, Pre-Mayan, Mayan, Anatolian, and Early Egyptian Data, with a Projection to 3000 AD,* and *Mankind in Decline: Genetic Disasters, Human-Animal Hybrids, Overpopulation, Pollution, Global Warming, Food and Water Shortages, Desertification, Poverty, Rising Violence, Genocide, Epidemics, Wars, Leadership Failure.*

He has taught approximately 4,000 students in undergraduate, graduate, and postgraduate corporate education courses primarily in major universities, and large companies and government agencies.

He developed a quantum theory, The Unified SuperStandard Theory (UST), which describes elementary particles in detail without the difficulties of conventional quantum field theory. He found that the internal symmetries of this theory could be exactly derived from an octonion theory called QUeST. He further found that another octonion theory (UTMOST) describes the Megaverse. It can hold QUeST universes such as our own universe. It has an internal symmetry structure which is a superset of the QUeST internal symmetries.

Recently he developed Octonion Cosmology. He replaced it with HyperCosmos theory, which has significantly better features. He developed a fractionalization process for dimensions, particles and symmetry groups. He also described transformation that reduced particles and dimensions to a far more compact form. He also developed a precursor theory ProtoCosmos that leads to the HyperCosmos.

The author showed that space-time and Internal Symmetries can be unified in any of the ten HyperCosmos spaces in their associated HyperUnification spaces. The combined set of HyperUnification spaces enable all HyperCosmos dimensions to be obtained by a General Relativistic transformation from one primordial dimension in the 42 space-time dimension unified HyperUnification space.

At present the author devel;oped the Cosmos Theory that incorporates ProtoCosmos Theory, HyperCosmos Theory, Limos Theory, Second Kind HyperCosmos Theory and HyperUnification Spaces. He has introduced PseudoFermion wave functions and theory, He has related Cosmos Theory to Regge trajectories of spaces, parton theory, Veneziano amplitudes, Fibonacci numbers and Ramsey numbers. He has calculated an approximation to the difficult R(n,n) Ramsey numbers.

He has developed a Gambol Model that successfully accounts for e-p deep inelastic scattering, fundamental particle resonances, hadron scattering, and the inner structure of particles based on confinement through Casimir forces of ideal gambol gases. The Gambol Planckian Distribution was derived.

He has applied the Gambol Model to particles, universes, and the Cosmos of universes. He showed that the Cosmos may have a distribution of 23 universes corresponding to various Cosmos spaces.

Recently he showed that Cosmos Theory follows from the number of independent asymmetric tensors in a dimension r. He also showed the close parallel between the form of γ-matrices and Cosmos Theory dimension arrays. The closeness suggested that dimension arrays have the same importance as γ-matrices for fermions.

He demonstrated that the pressure of fermions within a space of dimension r balances the Casimir vacuum energy force for 18 dimensions. He showed that $2e\pi = 17.02$ marks the critical point where pressure balances Casimir force, which implies $r = 18$ is the highest dimension Physical Cosmos space. The dimension $2e\pi$ appears to set the approximate dimension for Cosmos spaces with dimension array size $2^{r+4} \cong (17.02/8)^{r+4}. \cong (e\pi/4)^{r+4}. \cong 2.13^{r+4}$.

9 798989 408443